THE PUZZLE
OF PAIN

THE
PUZZLE
OF PAIN

RONALD MELZACK

Basic Books, Inc., Publishers

NEW YORK

To the memory of Harry Handel

© 1973 by Ronald Melzack
Library of Congress Catalog Card Number: 73–81726
SBN 0–465–09521–6
Manufactured in the United States of America
10 9 8 7

Contents

Foreword

Intense, chronic pain has a profound, distressing influence on man's physical and mental well being. Man has therefore developed elaborate ways of ridding himself of pain. Most of them to date have been relatively unsuccessful.

Professor Melzack in this book has done more than rearrange the pieces of the "puzzle of pain." He has given us a logical development of the concepts of pain, both psychological and physiological, bringing us up to date with the advent of the "gate-control" theory which he and Professor Patrick D. Wall proposed in 1965. The theory provided a new thrust in neuro-physiologic thinking, and was quickly translated into a thera-peutic reality for the relief of pain by direct stimulation of nerves through the skin or of the central nervous system using highly sophisticated electronic devices. Although there is still controversy over the exact details of the "gate-control" theory, both Melzack and Wall have modified their original concept in the face of new data. Few theories in neurophysiology have generated so much heat and, hopefully, there will be additional light as facts accumulate from the laboratory and clinic.

This is a rare instance where pure science has had an imme-diate impact on the human situation, namely, new ways of thinking and dealing with pain. Melzack has written this book in such a way that it will be useful at all levels of interest from the lay person to the professional, and the publishers are to be congratulated for introducing what I believe to be a new modern medical classic.

No one can escape some pain in his lifetime, and we should remember the admonition of Leriche:

"Not everyone has a soul of fire, and in actual human life, even in the case of the great mystics, the struggle against pain exacts a high price."

Blaine S. Nashold, Jr., M.D.
Associate Professor of Neurosurgery, Department of Surgery,
Duke University Medical Center, Durham

Preface

Pain is one of the most important concerns of man. Acute pain serves the useful purpose of warning the individual of something wrong and also serves as a useful diagnostic aid for the physician. However, in its chronic pathologic form, pain imposes severe emotional and physical stresses on the patient and his family. Because pain is the most common disabling disease, it imposes severe economic stresses on the patient and society and thus constitutes a serious national and world health problem. Although accurate statistics are not available, data from a variety of sources suggest that chronic pain states cost the American people between 10 and 25 billions of dollars annually.

Even more important is the cost in terms of human suffering. It is a distressing fact that in this age of marvelous scientific and technologic advances which permit us to send people to the moon, there are still hundreds of thousands – indeed, millions – of suffering patients who are not getting the relief they deserve. Many of these patients are exposed to a high risk of complications from improper therapy, including narcotic addiction, or are subjected to multiple, often useless, and at times mutilating operations. A significant number give up medical care and consult quacks who not only deplete the patient's financial resources, but often do harm. Some patients with severe, intractable pain become so desperate as to commit suicide.

Despite the magnitude of the problem, medical students and physicians rarely receive organized training in the management of patients with chronic pain. Each specialist views pain in a narrow, tubular fashion, and many practitioners are unable or unwilling to devote the time and effort necessary for the care of these patients.

During the past decade there have been serious efforts to rectify these deficiencies in the management of pain. There has been an impressive surge of interest, among some basic scientists, in studying the mechanisms of chronic pain syndromes and in collaborating with clinical investigators and practitioners to solve some of the major problems. Outstanding among these has been the author of this book. Professor Ronald Melzack is one of the world's most respected authorities on pain because of his many critically important contributions to the field. These include extensive experimental studies of the neurophysiologic mechanisms and psychologic aspects of pain, and their complex interactions that give rise to pain experience in man. In collaboration with Professor Patrick D. Wall, he formulated the gate-control theory of pain which has attracted the attention of scientists throughout the world and is considered the most important recent advance in the field of pain. Melzack's studies and the gate-control theory have prompted many others to seek new approaches to the study of pain phenomena and have stimulated development of new methods to control chronic pain states.

In this book Professor Melzack provides a comprehensive summary of the most recent advances in the study and treatment of pain. The responsibility of the writer of a book of this nature is to evaluate, distill, and interpret the vast amount of available information and present it lucidly and concisely. Professor Melzack has met this challenge in a superb fashion. The book reflects not only his intense interest in the field but his critical analysis based on impressive clinical and research experience and his unusual talent and skill to communicate. It takes a giant step toward rectifying some of the deficiencies in pain management by providing current information essential for the proper care of patients in pain. I am confident that it will prove to be an effective tool for a better understanding of pain among laymen, medical students, and practicing physicians. The ultimate benefit will be improved patient care which is, after all, the crowning achievement of the medical scientist.

John J. Bonica, M.D.
Chairman, Department of Anesthesiology,
University of Washington School of Medicine, Seattle

Author's Note

The purpose of this book is to introduce both the student and the intelligent layman to the problem of pain – one of the most puzzling and challenging problems in biology and medicine. Pain is such a common experience that we rarely pause to define it in ordinary conversation. Yet no one who has worked on the problem of pain has ever been able to give it a definition which is satisfactory to all of his colleagues. Pain has obvious sensory qualities, but it also has emotional and motivational properties. It is usually caused by intense, noxious stimulation, yet it sometimes occurs spontaneously without apparent cause. It normally signals physical injury, but it sometimes fails to occur even when extensive areas of the body have been seriously injured; at other times it persists after all the injured tissues have healed and becomes a crippling problem that may require urgent, radical treatment.

There are many facets to the puzzle of pain. It has been studied in the laboratory by psychologists, physiologists, anatomists and pharmacologists. It has been examined in hospital clinics by neurologists, neurosurgeons, anaesthesiologists and specialists in internal medicine. Each of these biological or medical approaches has made a unique contribution towards understanding pain mechanisms. But the various approaches have given rise to conflicting observations and interpretations. Because every aspect of pain is the subject of vigorous debate, it is impossible to discuss pain without taking a theoretical point of view. As we shall see, a seemingly innocuous phrase such as 'pain receptors' presupposes a specific theoretical position. This book, therefore, examines the various facets of pain from a well-defined theoretical framework. The book contains two major sections. The first section (chapters 1–4) describes the psychological, clinical and physiological aspects of pain. The second section (chapters 5–7) examines the major theories of pain in terms of their ability to explain pain phenomena and their implications for the control of pain. The theoretical framework of the book, then, is implicit in the first section and is explicitly described in the second section. A glossary of basic medical terms is provided on p. 205.

I am grateful to many colleagues and friends who have helped guide me in my attempts to understand pain. D. O. Hebb introduced me to the problem and provided me with an exciting, new conceptual approach. W. K. Livingston led me through the subtleties and complexities of the problem, and his thinking has influenced every aspect of my subsequent work. In recent years, I have been deeply influenced and stimulated by P. D. Wall, whose research and ideas have had a powerful impact on the development of new concepts of pain; our joint explorations of new theoretical approaches and their implications are described throughout the pages of this book.

It gives me pleasure to thank Dalbir Bindra and Jane Stewart for their many excellent suggestions which have improved this book, and Mrs Jeannette Nevile for her outstanding secretarial assistance. I am also grateful to my wife Lucy and our children Laurie and Joey for their constant encouragement and support.

THE PUZZLE
OF PAIN

1 The Puzzle of Pain

Anyone who has suffered prolonged, severe pain comes to regard it as an evil, punishing affliction. Yet everyone recognizes the beneficial aspect of pain. It warns us that something biologically harmful is happening to our bodies.

Congenital insensitivity to pain

People who are born without the ability to feel pain provide convincing testimony on the value of pain (Sternbach, 1968). Many of these people sustain extensive burns, bruises and lacerations during childhood, frequently bite deep into the tongue while chewing food, and learn only with difficulty to avoid inflicting severe wounds on themselves. The failure to feel pain after a ruptured appendix, which is normally accompanied by severe abdominal pain, led to near death in one such man. Another man walked on a leg with a cracked bone until it broke completely. Children who are congenitally insensitive to pain sometimes pull out their own teeth and have on occasion even pushed their eyeballs out of their sockets (Jewesbury, 1951).

The best documented of all cases of congenital insensitivity to pain is Miss C., a young Canadian girl who was a student at McGill University in Montreal. Her father, a physician in Western Canada, was fully aware of her problem and alerted his colleagues in Montreal to examine her. The young lady was highly intelligent and seemed normal in every way except that she had never felt pain. As a child, she had bitten off the tip of her tongue while chewing food, and had suffered third-degree burns after kneeling on a hot radiator to look out of the window. When examined by a psychologist (McMurray, 1950) in the laboratory, she reported that she did not feel pain when noxious stimuli were presented. She felt no pain when parts of her body were subjected to strong electric shock, to hot water at temperatures that usually produce reports of burning pain, or to a prolonged ice-bath. Equally astonishing was the fact that she showed no changes in blood pressure, heart rate, or respiration when these stimuli were presented. Furthermore, she could not remember ever sneezing or coughing, the gag reflex could be elicited only with great difficulty, and corneal reflexes (to protect the eyes) were absent. A variety of other stimuli, such as inserting a stick up through the nostrils, pinching tendons, or injections of histamine under the skin – which are normally considered as forms of torture – also failed to produce pain.

Miss C. had severe medical problems. She exhibited pathological changes in her knees, hip and spine, and underwent several orthopaedic operations. Her surgeon attributed these changes to the lack of protection to joints usually given by pain sensation. She apparently failed to shift her weight when standing, to turn over in her sleep, or to avoid certain postures, which normally prevent inflammation of joints.

Miss C. died at the age of twenty-nine of massive infections that could not be brought under control. During her last month, she complained of discomfort, tenderness and pain in the left hip. The pain was relieved by analgesic tablets. There is little doubt that her inability to feel pain until the final month of her life led to the 'extensive skin and bone trauma that contributed in a direct fashion to her death' (Baxter and Olszewski, 1960, p. 381).

Spontaneous pain

In contrast to people who are incapable of feeling pain are those who suffer severe pain in the absence of any apparent stimulation. Damage of peripheral nerves in the arms or legs, by gunshot wounds or other injuries, is sometimes accompanied by excruciating pains that persist long after the tissues have healed and the nerve fibres have regenerated. These pains may occur spontaneously for no apparent reason. They have many qualities, and may be described as burning, cramping or shooting. Sometimes they are triggered by innocuous stimuli such as gentle touches or even a puff of air. Spontaneous attacks of pain may take minutes or hours to subside, but may occur repeatedly each day for years after the injury. The frequency and intensity of the spontaneous pain-attacks may increase over the years, and the pain may even spread to distant areas of the body. The initial cause of these pains is sometimes far more subtle than peripheral nerve damage. Minor injuries may give rise to astonishingly severe pain. In these cases, Livingston (1943, p. 110) notes:

The onset of symptoms may follow the most commonplace of injuries. A bruise, a superficial cut, the prick of a thorn or a broken chicken-bone, a sprain or even a post-operative scar may act as the causative lesion. The event which precipitates the syndrome may appear both to the patient and the physician as of minor consequence, and both have every reason to anticipate the same prompt recovery that follows similar injuries. This anticipation is not realized and the symptoms tend to become progressively worse.

One of these cases was described by Livingston (1943, p. 109):

Mrs G. E. A., aged fifty-eight, was referred for treatment of periodic pain in her right foot on 9 September 1937. Three years previously she fell and injured this foot. The outer side of the foot at the base of the toes turned 'black and blue', but X-ray plates did not reveal any fractures. As the ecchymosis (discolouration) cleared she noted that the outer three toes 'felt dead'. Later she began to have periodic pains 'like a toothache' in these toes and during such attacks all three would be extremely sensitive to touch. The attacks continued with increasing frequency and severity, sometimes occurring several times a day, and occasionally skipping a day or two, but never longer. . . . The subjective feeling of 'deadness' seemed to increase just before an attack began – next she experienced a sensation of swelling in the toes beginning at their bases on the plantar (sole) surface and spreading to involve all three toes to their junction with the foot. At its height she said the toes felt 'as if bursting and on fire'. During the attack she was unable to tolerate the lightest touch to the toes. She had never noted any change in colour, temperature or sweating of these toes even during an attack. For the previous two years the outer three toes of the other foot had felt 'slightly numb and dead' and on several occasions she had experienced twinges of pain in them which made her fear that 'the trouble is going over into the other foot'.

Nothing of significance was found in physical examination. She received ten injections of a 2 per cent novocaine solution into the (bottom) of the foot at the base of the toes. Each injection was followed by a period of complete relief from attacks, and these intervals of freedom became increasingly long as the treatment progressed. The last injection was given on 18 March 1938, since which time there has been no recurrence of pain.

Not all cases end as happily as this one. Sometimes the pain persists and becomes so unbearable that the patient may undergo successive surgical operations in the attempt to abolish it. In cases such as these, the pain serves no biologically useful purpose. It is as though some normally adaptive mechanism has run amok and, like the dangerous criminal whose mind may be brilliant but warped, needs to be isolated, contained and treated. Leriche (1939, p. 23), a brilliant surgeon who spent much of his life attempting to relieve suffering, contemplated this aspect of pain:

Defence reaction? Fortunate warning? But as a matter of fact, the majority of diseases, even the most serious, attack us without warning. When pain develops . . . it is too late . . . . The pain has only made more distressing and more sad a situation already long lost . . . . In fact, pain is always a baleful gift, which reduces the subject of it, and makes him more ill than he would be without it.

Buytendijk (1962, p. 40), who studied the psychology of pain, also concluded that pain 'is not merely a problem, but a mystery . . . a senseless element of life. It is a "malum" placed in opposition to life, an obstacle and a threat, which throws man aside like some wretched creature who dies a thousand times over again.'

The puzzle

These two kinds of cases – the inability to feel pain in spite of injury and spontaneous pain in the absence of injurious stimulation – represent the extremes of the full spectrum of pain phenomena. We do not yet have a satisfactory explana-

tion for either type of case. Instead we must resort to speculation and theory: the best possible guess on the basis of the
available evidence.

It was once thought that the mechanisms that subserve pain
would be entirely revealed if we applied noxious stimuli to the
skin and then mapped the pathways taken by nerve impulses
through the spinal cord and brain. Unfortunately, pain
mechanisms are not as simple as this. When the skin is
pinched or crushed, for example, it is true that receptors with
very high thresholds are stimulated, but so are receptors with
much lower thresholds which are ordinarily activated by
gentle touch or vibration. The same is true for extreme heat, or
cold, or any other noxious stimulus. Painful stimuli, in other
words, are usually extremes of other natural stimuli, and they
tend to activate receptors that may also be involved in eliciting
other sensations such as tickle, touch, warmth or cold. A
noxious stimulus, moreover, brings about a variety of other
changes, such as increased sweating and blood flow at the
skin, and these too would contribute to the afferent (sensory)
information going to the brain. How, then, is the neuro-
physiologist to know which portion of the afferent pattern is
related to pain? Or is it all related to pain? The critical
question is this: does the brain examine just a specific message
ascending along specific fibres, or does it monitor *all* the
input and make a decision in terms of the sheer number of
nerve impulses in all active fibres?

The answer to this question represents the key to the puzzle
of pain. It therefore has profound implications for its treatment. It was long hoped that we need only find the pathways
in the nervous system that send pain messages from the body
to the brain, and pain could be eliminated simply by cutting
the pathways. There are many forms of pain, however, that
defy this simple solution. Attempts to stop spontaneous pains
by cutting pathways in the spinal cord or the brain produce as
many failures as successes (Sunderland, 1968). Other kinds of
pain are more amenable to surgical treatment. Pain produced
by cancer is totally relieved by spinal cord surgery in about
50 per cent of patients, and is partially relieved in another 25

per cent. But the remainder – about one out of four – continue to suffer (Nathan, 1963). Even those who are helped sometimes report that they now have intense 'girdle pains' at the level of the operation – pains which they did not have before (Noordenbos, 1959). In a few cases, the pain after surgery may be worse than the pain for which the patients were treated (Drake and McKenzie, 1953).

An important aim of pain research is the successful treatment of pathological pain. The clinical syndromes which result from peripheral nerve injury bewilder the scientist who tries to understand them. Still worse, the failure to solve the problems they present means prolonged suffering and tragedy to many patients. People who face death due to a malignant disease such as cancer also face the prospect of extreme pain. Those who sustain brain damage as a result of a stroke may suffer severe pain (often called 'central pain') for the rest of their lives. Drugs help up to a point. Surgery offers some slight hope. But the pain may continue unabated until the end. Pain, then, is more than an intriguing puzzle. It is a terrible problem that faces all humanity and urgently demands a solution.

The field of pain research and theory has developed rapidly in recent years. These developments have come from many disciplines, including psychology, physiology and clinical medicine. As a result of this progress, exciting new techniques have been proposed for the treatment of pain. The purpose of this book is to describe the research and the theories, as well as the pursuit of new directions aimed at the control of pain.

2 The Psychology of Pain

The obvious biological value of pain as a signal of tissue damage leads most of us to expect that it must always occur after injury and that the intensity of pain we feel is proportional to the extent of the damage. Actually, in higher species at least, there is much evidence that pain is not simply a function of the amount of bodily damage alone. Rather, the amount and quality of pain we feel are also determined by our previous experiences and how well we remember them, by our ability to understand the cause of the pain and to grasp its consequences. Even the culture in which we have been brought up plays an essential role in how we feel and respond to pain.

When compared with vision or hearing, the perception of pain seems simple, urgent and primitive. We expect the nerve signals evoked by injury to 'get through', unless we are unconscious or anaesthetized. But experiments show that pain is not always perceived after injury even when we are fully conscious and alert. Thus a knowledge of pain perception goes beyond the problem of injury and the sensory signals of pain. The study of pain perception can help us to understand the enormous plasticity of the nervous system and the individual differences that make each of us respond to the world in a unique fashion (Melzack, 1961).

A vast amount of study has been devoted to the perception of pain, especially in the last decade, and from it is emerging a concept of pain that is quite different from the older views about pain. The evidence shows that pain is much more variable and modifiable than many people have believed in the past. Pain differs from person to person, culture to culture. Stimuli that produce intolerable pain in one person may be tolerated without a whimper by another. Indeed, masochists are known to seek out and enjoy injurious stimulation such as whipping or burning the skin. In some cultures, moreover, initiation rites and other rituals involve procedures that we associate with pain, yet observers report that these people appear to feel little or no pain. Pain perception, then, cannot be defined simply in terms of particular kinds of stimuli. Rather, it is a highly personal experience, depending on cultural learning, the meaning of the situation, and other factors that are unique to each individual.

Cultural determinants

Cultural values are known to play an important role in the way a person perceives and responds to pain. In Western culture, for example, childbirth is considered by many to be one of the worst pains a human being can undergo. Yet anthropologists (Kroeber, 1948) have observed cultures throughout the world which practise *couvade*, in which the women show virtually no distress during childbirth. In some of these cultures a woman who is going to give birth continues to work in the fields until the child is just about to be born. Her husband then gets into bed and groans as though he were in great pain while she bears the child. In more extreme cases, the husband stays in bed with the baby to recover from the terrible ordeal, and the mother almost immediately returns to attend to the crops.

Can this mean that all women in our culture are making up their pain? Not at all. It happens to be part of our culture to

Figure 1 The annual hook-swinging ceremony practised in remote Indian villages. *Top* shows two steel hooks thrust into the small of the back of the 'celebrant', who is decked with garlands. The celebrant is later taken to a special cart which has upright timbers and a cross-beam. *Bottom* shows the celebrant hanging on to the ropes as the cart is moved to each village. After he blesses each child and farm field in a village, he swings free, suspended only by the hooks. The crowds cheer at each swing. The celebrant, during the ceremony, is in a state of exaltation and shows no sign of pain.
(from Kosambi, 1967, p. 105)

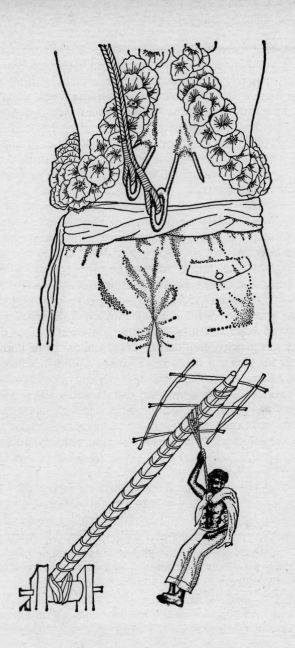

recognize childbirth as possibly endangering the life of the mother, and young girls learn to fear it in the course of growing up. Books (Dick-Read, 1962) on 'natural childbirth' ('childbirth without fear') stress the extent to which culturally determined fear enhances the amount of pain felt during labour and birth and point out how difficult it is to dispel it.

One of the most striking examples of the impact of cultural values on pain is the hook-hanging ritual still in practice in parts of India (Kosambi, 1967). The ceremony derives from an ancient practice in which a member of a social group is chosen to represent the power of the gods. The role of the chosen man (or 'celebrant') is to bless the children and crops in a series of neighbouring villages during a particular period of the year. What is remarkable about the ritual is that steel hooks, which are attached by strong ropes to the top of a special cart, are shoved under his skin and muscles on both sides of the back (Figure 1). The cart is then moved from village to village. Usually the man hangs on to the ropes as the cart is moved about. But at the climax of the ceremony in each village, he swings free, hanging only from the hooks embedded in his back, to bless the children and crops. Astonishingly, there is no evidence that the man is in pain during the ritual; rather, he appears to be in a 'state of exaltation'. When the hooks are later removed, the wounds heal rapidly without any medical treatment other than the application of wood ash. Two weeks later the marks on his back are scarcely visible.

Psychophysical studies

It is often supposed that variations in pain perception from person to person are due to different 'pain thresholds'. That is, people are assumed to be physiologically different from one another so that one person may have a low threshold (and feel pain after slight injury), while another has a high threshold (and feels pain only after intense injury). There is now evidence that all people, regardless of cultural background, have a uniform *sensation threshold* – that is, the lowest stimulus value at which sensation is first reported. Sternbach and Tursky (1965, p. 241) made careful measurements of sensation threshold, using electric shock as the stimulus, in American born women belonging to four different ethnic groups: Italian, Jewish, Irish, and Old American. They found no differences among the groups in the level of shock that was first reported as producing a detectable sensation. The sensory conducting apparatus, in other words, appears to be essentially similar in all people so that a given critical level of input always elicits a sensation.

This observation of a uniform sensation threshold is found only in precisely controlled laboratory experiments in which all environmental conditions are kept constant. In more natural situations, as we shall soon see, attention, suggestion, or other cognitive processes can radically modify the sensation threshold. A football player may not feel a severe kick on the shin during the excitement of the game, while an over-anxious, tense person may report that he feels severe pain when he is stimulated with very mild electric shocks.

Even in laboratory conditions, however, cultural background may have a powerful effect on the *pain perception threshold* – that is, the lowest stimulus level at which a person reports feeling pain. For example, levels of radiant heat that are reported as painful by people of Mediterranean origin (such as Italians and Jews) are described merely as warmth by Northern Europeans (Hardy, Wolff and Goodell, 1952). The most striking effect of cultural background, however, is on *pain tolerance levels*. Sternbach and Tursky (1965) report that the levels at which subjects refuse to tolerate electric shock, even when they are encouraged by the experimenters, depend, in part at least, on their ethnic origin. Women of Italian descent tolerate less shock than women of Old American or Jewish origin. In a similar experiment (Lambert, Libman and Poser, 1960), in which Jewish and Protestant women served as subjects, the Jewish, but not the Protestant, women increased their tolerance levels after they were told that their religious group tolerated pain more poorly than others.

These differences in pain tolerance reflect different ethnic attitudes towards pain. Zborowski (1952) found that Old

Americans have an accepting, matter-of-fact attitude towards pain and pain-expression. They tend to withdraw when the pain is intense, and cry out or moan only when they are alone. Jews and Italians, on the other hand, tend to be vociferous in their complaints and openly seek support and sympathy. The underlying attitudes of the two groups, however, appear to be different. Jews tend to be concerned about the meaning and implications of the pain, while Italians usually express a desire for immediate pain relief.

Just how far people can push themselves to tolerate pain is indicated by the Sun Dance 'self-torture' ceremonies of the North American Plains Indians. Each young man who participated in the ceremonies around the sacred Sun Dance pole first had two incisions made with a sharp knife on each side of his chest. Skewers were then inserted through one of the incisions, pushed under the skin and out of the other incision. Wissler (1921, p. 264), who observed the ceremony, describes the subsequent events:

This being done to each breast, with a single skewer for each, strong enough to tear away the flesh, and long enough to hold the lariats fastened to the top of the sacred pole, a double incision was made on the back of the left shoulder, to the skewer of which was fastened an Indian drum. The work being pronounced good by the persons engaged in the operation, the young man arose, and one of the operators fastened the lariats (to the skewers) giving them two or three jerks to bring them into position.

The young man went up to the sacred pole, and while his countenance was exceedingly pale, and his frame trembling with emotion, threw his arms around it, and prayed earnestly for strength to pass successfully through the trying ordeal. His prayer ended, he moved backward until the flesh was fully extended, and placing a small bone whistle in his mouth, he blew continuously upon it a series of short sharp sounds, while he threw himself backward, and danced until the flesh gave way and he fell. Previous to his tearing himself free from the lariats, he seized the drum with both hands and with a sudden pull tore the flesh on his back, dashing the drum to the ground amid the applause of the people. As he lay on the ground, the operators examined his wounds, cut off the flesh that was hanging loosely, and the ceremony was at an end. In former years the head of a buffalo was fastened by a rope on the back of the person undergoing the feat of self-immolation, but now a drum is used for that purpose.

From two to five persons undergo this torture every Sun Dance. Its object is military and religious. It admits the young man into the noble band of warriors whereby he gains the esteem of his fellows, and opens up the path to fortune and fame. Of course the applause of the people and the exhibition of courage are important factors in this rite, but its chief feature is a religious one.

In contrast to these studies, which demonstrate *variability* in pain tolerance, other psychophysical experiments are aimed at revealing a mathematically precise relationship between the measured stimulus input and the intensity of sensation reported by the subject. Stevens, Carton and Shickman (1958) asked subjects to estimate the magnitudes of a series of electric shocks of varying intensity by assigning a number to each that expressed the subjective intensity of the shock. They found that the stimulus–sensation relationship is best described as a mathematical power function, a fact which has been confirmed by several other investigators. The actual value of the exponent, however, varies from study to study (Sternbach and Tursky, 1964).

Psychophysical studies that find a mathematical relationship between stimulus intensity and pain intensity are often cited (Beecher, 1959; Morgan, 1961) as supporting evidence for the assumption that pain is a primary sensation subserved by a direct communication system from skin receptors to pain centre. A simple psychophysical function, however, does not necessarily reflect equally simple neural mechanisms. Activities in the central nervous system, such as memories of earlier cultural experience, may intervene between stimulus and sensation and invalidate any simple psychophysical 'law'. The use of laboratory conditions that minimize such activities or prevent them from ever coming into play reduces the functions of the nervous system to those of a fixed-gain transmission line. It is under these conditions that psychophysical functions prevail.

Past experience

The evidence that pain is influenced by cultural factors leads naturally to an examination of the role of early experience in adult behaviour related to pain. It is commonly accepted that children are deeply influenced by the attitudes of their parents towards pain. Some families make a great fuss about ordinary cuts and bruises, while others tend to show little sympathy towards even fairly serious injuries. There is reason to believe, on the basis of everyday observations, that attitudes towards pain acquired early in life are carried on into adulthood.

The influence of early experience on the perception of pain has also been demonstrated experimentally. Melzack and Scott (1957) raised Scottish terriers in isolation cages from infancy to maturity so that they were deprived of normal environmental stimuli, including the bodily knocks and scrapes that young animals get in the course of growing up. They were surprised to find that these dogs, at maturity, failed to respond normally to a variety of noxious stimuli. Many of them poked their noses into a flaming match, withdrew reflexively, and then immediately sniffed at the flame again. If they snuffed out the flame, they reacted similarly to a second flaming match and even to a third. Others did not sniff at the match but made no effort to get away when the experimenters touched their noses with the flame repeatedly. These dogs also endured pin pricks with little or no evidence of pain. They invariably oriented to the pin as it penetrated the skin, but there was an absence of strong emotional arousal or behavioural withdrawal. In contrast, the litter-mates of these dogs that had been reared in a normal environment recognized potential harm so quickly that the experimenters were usually unable to touch them with the flame or pin more than once.

This astonishing behaviour of dogs reared in isolation cannot be attributed to a general failure of the sensory conducting systems. Intense electric shock elicited strong emotional excitement and the dogs made obvious attempts to escape from it. Moreover, reflex movements made by the dogs during contact with fire and pin prick indicate that they felt something during stimulation; but the lack of observable emotional disturbance, apart from reflex or orienting movements, suggests that their *perception* of actual damage to the skin was highly abnormal.

This abnormal behaviour may be due, in part at least, to a failure to attend selectively to noxious stimuli when they are presented in an unfamiliar environment in which all stimuli are equally attention-demanding. It is apparent (Melzack, 1965, 1969) that young organisms learn which environmental stimuli are important and which are not. The dogs reared in isolation had no way of knowing that small objects, such as flaming matches or pins, are more important than large objects such as chairs or tables. In the absence of such information, normally acquired early in life, all stimuli are equally relevant (or irrelevant). As a result, the dogs would have difficulty differentiating and responding adaptively to stimuli, including those that are potentially damaging. The results suggest, therefore, that the significance – or meaning – of environmental stimuli acquired during early experience plays an important role in pain perception. It is important to note that heredity may determine the extent to which early experience influences later behaviour. Beagles raised in isolation cages are not as severely disturbed as Scotties or mongrels and are capable of behaving more normally towards flaming matches and pin pricks (Lessac, 1965).

Meaning of the situation

There is considerable evidence to show that people also attach variable meaning to pain-producing situations and that the meaning greatly influences the degree and quality of pain they feel. During the Second World War, Beecher (1959) observed the behaviour of soldiers severely wounded in battle. He was astonished to find that when the wounded men were carried into combat hospitals, only one out of three complained of enough pain to require morphine. Most of the soldiers either denied having pain from their extensive wounds or had so little that they did not want any medication to relieve it. These men, Beecher points out, were not in a state of shock, nor were they totally unable to feel pain, for they complained as vigorously as normal men at an inept vein puncture. When Beecher returned to clinical practice after the war, he asked a group of civilians who had surgical wounds similar to those received by the soldiers whether they wanted morphine to alleviate their pain. In contrast with the wounded soldiers, four out of five claimed they were in severe pain and pleaded for a morphine injection.

Beecher (1959, p. 165) concluded from his study that:

The common belief that wounds are inevitably associated with pain, and that the more extensive the wound the worse the pain, was not supported by observations made as carefully as possible in the combat zone . . . . The data state in numerical terms what is known to all thoughtful clinical observers: there is no simple direct relationship between the wound *per se* and the pain experienced. The pain is in very large part determined by other factors, and of great importance here is the significance of the wound . . . . In the wounded soldier (the response to injury) was relief, thankfulness at his escape alive from the battlefield, even euphoria; to the civilian, his major surgery was a depressing, calamitous event.

The importance of the meaning associated with a pain-producing situation is made particularly clear in conditioning experiments carried out by Pavlov (1927, 1928). Dogs normally react violently when they are given strong electric shocks to one of the paws. Pavlov found, however, that if he consistently presented food to a dog after each shock, the dog developed an entirely new response. Immediately after a shock the dog would salivate, wag its tail and turn eagerly towards the food dish. The electric shock now failed to evoke any responses indicative of pain and became instead a signal meaning that food was on the way. This type of conditioned behaviour was observed as long as the same paw was shocked. If the shocks were applied to another paw, the dogs reacted violently. Pavlov reports that similar results were obtained in other experiments in which intense pressure or heat were used as the conditioned stimuli. This study shows convincingly that stimulation of the skin is localized, identified and evaluated *before* it produces perceptual experience and overt behaviour.

The meaning of the stimulus acquired during earlier conditioning modulates the sensory input before it activates
brain processes that underlie perception and response.

There are more familiar examples of the role played by
personal evaluation of the situation. A slap on the behind,
when administered to a child in the course of play, may be
ignored or produce laughter. The same slap, given in another
context, such as punishment for a misdeed, may produce tears
and howls of terrible pain. Similarly, abdominal sensations
that are assumed to be gas cramps and are usually ignored
may be felt as severe pain after learning that a friend or
relative has stomach cancer. The pain may persist and get
worse until a doctor assures the person that nothing is wrong.
It may then vanish suddenly. Still another example is the
frequent observation by dentists that patients who arrive early
in the morning, complaining of a terrible toothache that kept
them awake all night, sometimes report that the pain disappeared when they entered the dentist's office. They may even
have difficulty remembering which tooth had hurt. The presence or absence of pain in these patients is clearly a function
of the meaning of the situation: the pain was unbearable
when help was unavailable, and diminished or vanished when
relief was at hand.

Attention, anxiety and suggestion

Attention to stimulation also contributes to the intensity of
pain experience. It is well known that boxers, football players
and other athletes can sustain severe injuries during the excitement of the sport without being aware that they have been
hurt. In fact, almost any situation that attracts a sufficient
degree of intense, prolonged attention may provide the
conditions for other stimulation to go by unnoticed, including
wounds that would cause considerable suffering under normal
circumstances.

The manipulation of attention, together with strong suggestion, are both part of the phenomenon of hypnosis. The
hypnotic state eludes precise definition. But, loosely speaking,
hypnosis is a trance state in which the subject's attention is

focused intensely on the hypnotist while attention to other stimuli is markedly diminished. After people are hypnotized they can, with appropriate suggestion, be cut or burned yet report that they did not feel pain (Barber, 1969). They may say that they felt a sharp tactile sensation or strong heat, but they maintain that the sensations never welled up into pain. Evidently a small percentage of people can be hypnotized deeply enough to undergo major surgery entirely without anaesthesia. For a larger number of people hypnosis reduces the amount of pain-killing drug required to produce successful analgesia.

Self-hypnosis or auto-suggestion may be related to the state of meditation observed in mystics or other profoundly religious people. Deep meditation, or prolonged, intense focusing of attention on inner feelings, thoughts or images, may produce a state similar to hypnotic analgesia. Indian fakirs have frequently been observed to walk across beds of hot coals, or lie on a bed of nails or cactus thorns without evidence of pain. It is possible that the fakirs develop highly calloused skin. But this cannot be the whole explanation. It is more likely that they enter a trance-like state as a result of deep meditation. (In other cultures, the same effect may be achieved by a prolonged period of singing, drumming and dancing.) The human ability to voluntarily direct attention towards inner feelings, thoughts or images, and to block out all extraneous environmental inputs may also explain the observations (Huxley, 1952) that men and women who were burned at the stake for their religious beliefs were sometimes observed to experience what can only be described as ecstasy (although others were certainly in agony). It is possible, of course, that the meaning of the situation also played an important role in the behaviour of these people. The prospect of certain salvation of the soul or an imminent meeting with their Maker may have contributed to the transformation of the somatic input evoked by the flames so that it evoked ecstasy rather than agony.

In contrast to the effects of distraction, if a person's attention is focused on a potentially painful experience, he will tend to perceive pain more intensely than he would normally. Hall and Stride (1954) found that the simple appearance of the word 'pain' in a set of instructions made anxious subjects report as painful a level of electric shock they did not regard as painful when the word was absent from the instructions. Thus the mere anticipation of pain is sufficient to raise the level of anxiety and thereby the intensity of perceived pain. Similarly, Hill, Kornetsky, Flanary and Wikler (1952a and b) have shown that if anxiety is dispelled (by reassuring a subject that he has control over the pain-producing stimulus), a given level of electric shock or burning heat is perceived as significantly less painful than the same stimulus under conditions of high anxiety. They were also able to show that morphine diminishes pain if the anxiety level is high but has no demonstrable effect if the subject's anxiety has been dispelled before the administration of the drug.

The influence of suggestion on the intensity of perceived pain is further demonstrated by studies of the effectiveness of placebos. Clinical investigators (Beecher, 1959) have found that severe pain, such as post-surgical pain, can be relieved in some patients by giving them a placebo (usually some non-analgesic substance such as a sugar or salt solution) in place of morphine or other analgesic drugs. About 35 per cent of the patients report marked relief of pain after being given a placebo. Since morphine, even in large doses, will relieve severe pain in only about 75 per cent of patients, one can conclude that nearly half of the drug's effectiveness is really a placebo effect.

Because suggestion, even in the subtlest form, may have a powerful effect on pain experience, the use of the 'double-blind' technique is essential in the evaluation of drugs. When this technique is used, both the experimental drug and the placebo are labelled in such a way that neither the patient nor the physician know which one has been administered. Only then can the effect of the drug be evaluated in comparison with a physiologically neutral control chemical agent. The remarkably powerful effect of a placebo in no way implies that people who are helped by a placebo do not have real pain; no one will deny the reality of post-surgical pain. Rather, it illustrates the powerful contribution of suggestion to the perception of pain.

It is not clear precisely how a placebo works. It is generally assumed that the suggestion itself is sufficient to produce the entire placebo effect. Indeed, strong suggestion alone is almost as effective as suggestion given after prolonged hypnotic induction in producing increased pain tolerance (Hilgard, 1965; Barber, 1969). However, the placebo may also decrease anxiety because it makes the patient believe that something is being done to relieve the pain. Both effects probably always occur together. Whatever the explanation, it is clear that the physician may often relieve pain significantly by prescribing placebos to influence cognitive processes as well as by treating the injured areas of the body.

Audio analgesia

The complexity of the psychological contributions to pain, and the difficulty encountered in examining them experimentally are both illustrated by the phenomenon known as 'audio analgesia'. The discovery by Gardner and Licklider (1959) that intense auditory stimulation (white noise) suppresses pain produced by dental drilling and extraction was named 'audio analgesia', and created great excitement as a prospective new tool for the control of pain in dentistry. The discovery led rapidly to the manufacture and sale of several impressive-looking models of a machine which relayed white noise and stereophonic music to the dental patient through earphones. The machine sold under a variety of trade-names and, in the fierce competition that ensued, components of increasingly high quality were used, and the price of the equipment rose accordingly.

The effects of the audio procedure, on some patients, were dramatic, and even tooth extractions were reported by the patients to be totally painless. The phenomenon, then, when it occurred, was striking – *but*, it did not always occur. The machine worked splendidly for some patients, not at all for others. Moreover, the fact that laboratory studies using radiant heat applied to the skin (Camp, Martin and Chapman, 1962) or electric shocks to the teeth (Carlin, Ward, Gershon and Ingraham, 1962) failed to demonstrate any effect of auditory stimulation on pain threshold suggested that the dramatic results reported in the clinic could not be attributed to any single, simple mechanism. Other variables such as suggestion, distraction and reduction of anxiety also had to be considered.

The critical dimension of rate of pain increase

Exploratory studies were carried out by Melzack, Weisz and Sprague (1963) in an attempt to uncover the mechanisms involved in audio analgesia. They found that most laboratory pains (such as electric shock or radiant heat) rose so suddenly and sharply to unbearable intensities that subjects were unable to control them by distraction or other means. However, when a radiant heat lamp was held sufficiently far from the subject's skin, the pain rose slowly enough to allow the subject time to use distraction or other methods in his attempt to control it.

It then became apparent that the amount of pain tolerated by the subjects was often determined by their expectation of future pain on the basis of rate of pain increase rather than by pain intensity level as such. Thus, many subjects expected the pain to continue to rise with constant slope and anticipated (by its temporal course) the point at which it would reach an intolerable level (Figure 2). Hence, they asked the experimenter to turn off the pain source, but frequently followed the request by saying, 'I'm sure I could have taken more, but I was afraid to go to the *real* limit.' When the subjects risked exposure to higher pain levels on subsequent trials, with the aid of auditory stimulation to distract attention or other stratagems such as strong auto-suggestion ('The pain isn't bad – I can take more'), they often found that the pain levelled off or decreased *before* it became intolerable (Figure 2). The stratagems, then, enabled the subjects to discover that their expectation was wrong. The subjects were subsequently less anxious, and were able to endure the pain for longer periods of time.

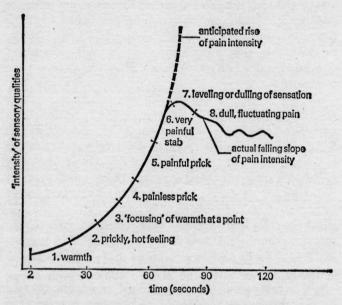

Figure 2 Idealized curve of sensory qualities produced by radiant heat, based on reports of subjective sensation (from Melzack, Weisz and Sprague, 1963, p. 239)

Contributions of auditory stimulation and suggestion

A direct study (Melzack, Weisz and Sprague, 1963) of 'audio analgesia' was carried out using the cold pressor test (immersion of the hand into an ice bath), which produces a deep pain that has even more slowly rising temporal properties than radiant heat pain. Auditory stimulation was provided by a modified white noise generator, a tape-deck with stereophonic music, stereophonic earphones, and a control box held by the subject's free hand. The control box had two intensity-control knobs, one for music, the other for noise. A *placebo stimulus* was supplied by the same apparatus, except that a switch prevented music or noise from reaching the subject's ears; instead he was provided with a low-intensity, sixty-cycle hum which increased slightly in volume when the noise-control knob was turned.

Three groups of subjects received, in alternation, two control sessions (without auditory stimulation or suggestion) to determine the duration that they could tolerate the pain, and two experimental sessions with the following conditions:

Subjects in *Group 1* received strong auditory stimulation, but no explicit suggestion about the purpose of the music and noise. Each subject received the music and noise at maximum intensity and was instructed to turn the noise-intensity knob to maintain a constant intensity relationship between the music and noise. The subject was instructed to say 'stop' when he experienced as much pain as he could stand. Subjects in *Group 2* received the same instructions, together with strong suggestion that intense auditory stimulation had been found by dentists across the country to be extremely effective for alleviating pain. The subject was 'informed' that 'intense sound prevented pain from reaching consciousness'. Subjects in *Group 3* were given the placebo stimulus with strong suggestion. They were told that ultrasonic sound was found by dentists across the country to be highly effective for alleviating pain since it prevented pain from reaching consciousness. Each subject was also told that he would hear a low-frequency hum whose volume would indicate the amount of ultrasonic sound to which he was being exposed. It was suggested that he increase the volume of the ultrasonic sound as the pain increased, 'since higher volumes produce more relief than low volumes'.

The results of the study were clear. Intense auditory stimulation accompanied by strong suggestion that it abolishes pain (Group 2) produced a marked increase in pain-tolerance duration compared with the control condition without auditory stimulation. In contrast, intense auditory stimulation without explicit suggestion (Group 1) or strong suggestion accompanied by the elaborate placebo stimulus but without intense auditory stimulation (Group 3) did not increase pain-tolerance durations compared with the control condition.

The manner in which the subjects employed the music and noise to 'control' their pains is of particular interest in interpreting the results of the study. It was evident that they did not just passively receive the auditory stimulation but concentrated on the music by tracking it with the noise-volume knob, keeping time by tapping their feet, singing out loud, and so forth. In short, they actively forced their attention away from the compelling, slowly rising pain. Auditory stimulation, then, did not 'abolish' pain, but was used as a tool to divert attention away from the pain. The term 'audio analgesia' was clearly a misnomer: the auditory input did not produce analgesia but, instead, served as a stratagem to modulate pain tolerance.

These data, gained by laboratory experiment, provided an insight into the mechanisms underlying 'audio analgesia'. They suggested that the device could be effective in the hands of dentists with strong personalities, who could suggest convincingly to their patients that they would feel no pain, but not in the hands of those who used the machines with trepidation or simply placed earphones over the patients' ears and began their operations. Moreover, there was indication from dentists that the patient's personality was an important variable: some people are more suggestible than others, and this variable, of course, would interact with the dentist's personality. It is not surprising, therefore, that the enthusiasm for 'audio analgesia' first exhibited by the dental profession fell rapidly, and the machines were soon relegated to the attic of dental history.

Cognitive control of pain

The story of 'audio analgesia' contains an important moral: there is no easy, magical relief for pain. 'Audio analgesia' fell to the wayside because it did not block pain in an all-or-none, mechanical fashion. If we look at pain more realistically, however, and recognize its multiple determinants, 'audio analgesia' did not, in fact, fail. The auditory input, as long as it was recognized as part of a complex situation involving suggestion as well as personality variables, was an effective stratagem for many subjects and enabled them to tolerate pain that they would otherwise have found to be unbearable.

Psychogenic pain

Because psychological factors play such a powerful role in pain, several clinical pain syndromes have been labelled as 'psychogenic', with the implication that the primary cause of the pain is psychological. That is, the person is presumed to be in pain because he needs or wants it. A typical case has been reported by Freeman and Watts (1950, pp. 354–5):

A woman of hysterical temperament began at the age of sixteen to complain of abdominal pain so persistently that she accumulated a series of twelve to eighteen abdominal operations, with what might be termed progressive evisceration. Following a trivial head injury, she complained so bitterly of pain in the head that a subtemporal decompression was performed. From 1934 to 1936 she was confined to bed because of agonizing pain in the back and limbs. Examination showed swollen knuckles, tender knee joints with contracture, and roentgenograms of the spine revealed lipping of the vertebrae. When we saw her for the first time, she appeared uneasy, would not give her history and began wincing and overbreathing before the bed covers were turned down. She lay constantly on her left side and cried out if any attempt was made to turn her on her back. She defended herself with her right hand from any examination of her back, and when the right hand was restrained and the region of the sacrum was gently stroked, she screamed and trembled violently. On account of exaggeration of the complaints with very little anatomic substrate, a diagnosis of conversion hysteria with polysurgical addiction was made.

The concluding sentence of this case history suggests that the patient suffered pain primarily because of psychological needs, and that she became addicted to multiple surgical operations as a way of satisfying her needs. This woman, Freeman and Watts (1950) report, then underwent a frontal lobotomy (to cut the neural connections between the frontal cortex and thalamus). The operation did not entirely relieve her pain, but she was not bothered by it as much and was able to live a useful life.

A sympathetic analysis of this kind of patient is presented by Sternbach (1970, p. 182):

Such patients may acquire thick hospital charts over the years, as they have one operation after another and make the rounds from doctor to doctor. It is tempting to label them 'hysterics' or 'crocks', but such labels do not help them nor add to our understanding. Naming, as we know, is not explaining. Now it turns out that psychiatric studies of such patients have found them to be depressed. The depression may result from loss, as in adequate grief reactions, or it may result from intropunitive reactions, as happens when anger is not appropriately directed outwardly. In chronic conditions such patients become 'pain-prone'. You must understand that these people are not faking or malingering; they have real pain by any measure we can devise, and their suffering is manifest. But neurological models do not adequately describe them, at least not yet as well as psychiatric ones do. The tactics of treating such patients are those that will relieve depression. This may consist of psychotherapy, which encourages a complete response of mourning or which enables the patient to learn that it is safe and appropriate to express anger directly. Or, if the patient is manipulating or controlling his family with his symptom, family therapy may be necessary to change the system. Or, if therapy is not appropriate for the particular patient, then antidepressant medications, or mood elevators, can be very successful. In fact, for these people with persistent or recurrent pain with no apparent organic basis, analgesics may only worsen their symptoms, whereas antidepressants do much to alleviate them.

It is clear that we must recognize the psychological contribution to pain, but we must maintain a balanced view of it. Psychological factors contribute to pain, and pain may be helped by using psychological approaches. But there are, as we shall see, other contributions. This does not deny the existence of patients who need their pain, and whose lives derive meaning from it. Such patients complain of terrible pain yet discontinue certain types of therapies because of minor unpleasantness, such as an injection or the taste of a particular drug. However, even when psychological factors appear to play a major role, there is often tissue pathology which can also be treated. In such cases, the physical as well as the psychological symptoms require treatment.

The language of pain

The description of pain is a daily concern to the practising physician. Yet few studies have attempted to specify the dimensions of pain experience. The methods currently used for pain measurement treat pain as though it were a single, unique quality that varies only in intensity. The most common of these methods is the use of words such as 'mild','moderate', and 'severe', and subjects (or patients) are asked to choose the word that best describes the intensity of their pain. Another method consists of a five-point scale which ranges from 1 = mild pain to 5 = unbearable pain, and subjects are asked to choose the most appropriate number. In this way, some quantitative measure of pain is obtained. Still another method is the use of fractions, so that subjects who have received injections of analgesic drugs such as morphine are asked whether their pain is $\frac{1}{3}$, $\frac{1}{2}$, or $\frac{2}{3}$ of what it was before the injection. These simple methods have all been used effectively in hospital clinics, and have provided valuable information about the relative effectiveness of different drugs.

All of these methods specify only intensity. It is clear, however, that to describe pain solely in terms of intensity is like specifying the visual world only in terms of light flux without regard to pattern, colour, texture, and the many other dimensions of visual experience.

Pain, we now believe, refers to a category of complex experiences, not to a specific sensation that varies only along a single intensity dimension. The word 'pain', in this formulation, is a linguistic label that categorizes an endless variety of qualities. There are the pains of a scalded hand, a stomach ulcer, a sprained ankle; there are headaches and toothaches. Each is characterized by unique qualities. The pain of a toothache is obviously different from a pin prick, just as the pain of a coronary occlusion is uniquely different from the pain of a broken leg.

Clinical investigators have long recognized the varieties of

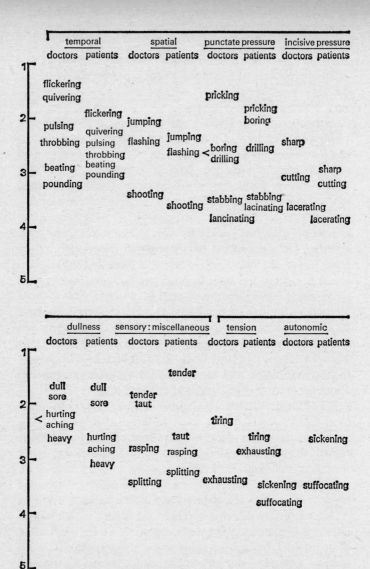

Figure 3 Spatial display of pain descriptors which have the same rank order, on an intensity scale, for doctors and patients. The scale values range from 1 (mild) to 5 (excruciating). Two words connected by an arrowhead have the same mean scale value. (from Melzack and Torgerson, 1971, p. 50)

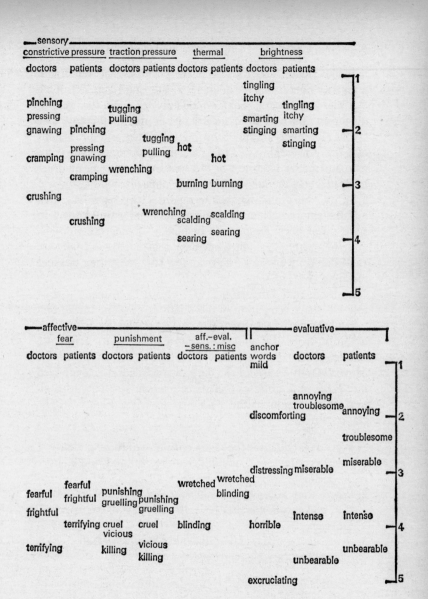

pain experience. Descriptions of the burning qualities of pain after peripheral nerve injury, or the stabbing, cramping qualities of visceral pains frequently provide the key to diagnosis and may even suggest the course of therapy. The layman is equally aware of the many qualities and dimensions of pain. An evening of radio, television or newspaper commercials makes us aware of the splitting, pounding qualities of headaches, the gnawing, nagging pain of rheumatism and arthritis, the cramping, heavy qualities of menstrual pain, and the smarting, itching qualities apparently well known to sufferers of piles. Despite the frequency of such descriptions, and the seemingly high agreement that such adjectives are valid descriptive words, there are few studies of their use and meaning.

Melzack and Torgerson (1971, p. 50) have made a start towards the specification of the qualities of pain. In the first part of their study, subjects were asked to classify 102 words, obtained from the clinical literature relating to pain, into smaller groups that describe different aspects of the experience of pain. On the basis of the data, the words were categorized into three major classes and sixteen subclasses. The distribution of a portion of the words is shown in Figure 3. The classes are:

1. Words that describe the *sensory qualities* of the experience in terms of temporal, spatial, pressure, thermal, and other properties.
2. Words that describe *affective qualities*, in terms of tension, fear, and autonomic properties that are part of the pain experience.
3. *Evaluative* words that describe the subjective overall intensity of the total pain experience.

Each subclass, which was given a descriptive label, consists of a group of words that were considered by most subjects to be qualitatively similar. Some of these words are undoubtedly synonyms, others seem to be synonymous but vary in intensity, while many provide subtle differences or nuances (despite their similarities) that may be of importance to a patient who is trying desperately to communicate to a physician.[1]

The second part of the study was an attempt to determine the pain intensities implied by the words within each subclass. Groups of doctors, patients, and students were asked to assign an intensity value to each word, using a numerical scale ranging from least (or mild) pain to worst (or excruciating) pain. When this was done, it was apparent that several words within each subclass had the same relative intensity relationships in all three sets. For example, in the spatial subclass, 'shooting' was found to represent more pain than 'flashing', which in turn implied more pain than 'jumping'. Although the precise intensity values differed for the three groups, all three agreed on the positions of the words relative to each other. The scale values of the words for doctors and patients are shown in Figure 3.

Because of the high degree of agreement on the intensity relationships among pain descriptors by subjects who have different cultural, socio-economic, and educational backgrounds, it has been possible to develop a questionnaire to determine the properties of different pain syndromes (Melzack, 1973). Such a questionnaire may one day contribute to diagnosis, and may also become a valuable experimental tool for studies of the effects of analgesic drugs and other methods of pain management.

Towards a definition of pain

The diversity of pain experiences explains why it has been impossible, so far, to achieve a satisfactory definition of pain. Pain is not a single quality of experience that can be specified in terms of defined stimulus conditions. It may be agreed that pain, like vision and hearing, is a complex perceptual experi-

1. In her essay 'On Being Ill' Virginia Woolf touches on precisely this point: 'English,' she writes, 'which can express the thoughts of Hamlet and the tragedy of Lear, has no words for the shiver and the headache. . . . The merest schoolgirl, when she falls in love, has Shakespeare and Keats to speak for her; but let a sufferer try to describe a pain in his head to a doctor and language at once runs dry.'

ence. But the many, diverse causes of pain prevent the specification of a particular kind of environmental energy as the stimulus for pain, in the way that light is the adequate stimulus for vision and air pressure waves for hearing. The word 'pain' represents a category of experiences, signifying a multitude of different, unique events having different causes, and characterized by different qualities varying along a number of sensory and affective dimensions.

It is obvious that many of the words in Figure 3 have two meanings. 'Burning', for example, can be used to refer to the sensation evoked when the skin is actually being burned. It can also be used by patients with peripheral nerve injury to describe their pain in the absence of any stimulus-evoked input. The word, in this case, has an analogy (or 'as if') meaning; the skin feels *as if* it were being burned. The use of the word in this case is not 'stimulus error'. Rather, it indicates that we tend to use words that have familiar, common meanings. A splitting headache, then, does not mean that the head is being split open. It obviously represents a figure of speech, meant to convey some property of the total pain experience – that the pain feels *as if* the head were being split open. In all of these cases, however, there is a somatic contribution to the experience. This contribution, as we shall see in chapter 4, can be highly complex. It includes afferent activity evoked by stimulation as well as more subtle, abnormal physiological activity in the central nervous system which may have been produced by earlier injury or even by chronic low-level irritation.

The analysis of the language of pain in the preceding section points the way towards a definition of pain. It suggests that pain may be defined in terms of a multidimensional space comprising several sensory and affective dimensions. The space comprises those subjective experiences which have both somatosensory and negative-affective components and that elicit behaviour aimed at stopping the conditions that produce them. If injury or any other noxious input fails to evoke negative affect and aversive drive (as in the cases described earlier of the football player, the soldier at the battlefront, or

Pavlov's dogs) the experience cannot be called pain. Conversely, anxiety or anguish without concomitant activity in the somatic afferent system is not pain. The 'pain' of bereavement or the 'heartache' of the scorned lover do not legitimately fall within this definition, although both psychological states may *contribute* to pain by modulating the afferent input.

The evaluative words in Figure 3 reflect the capacity of the brain to evaluate the importance or urgency of the overall situation. These words represent judgements based not only on sensory and affective qualities, but also on previous experiences, capacity to judge outcome, and the meaning of the situation. Thus, by reflecting the total circumstances at a given time, they serve to locate the position of the pain experience within the multidimensional space for the particular individual.

Implications of the psychological evidence

Taken together, the psychological data refute the concept that the intensity of noxious stimulation and the intensity of perceived pain have a one-to-one relationship. A stimulus may be painful in one situation and not in another. The same injury can have different effects on different people or even on the same person at different times. The data indicate that psychological variables may intervene between stimulus and perception and produce a high degree of variability between the two. In most instances, to be sure, a simple relationship holds: the harder the slam of a hammer on the thumb, the greater the pain is likely to be. The exceptions, however, illuminate the nature of the underlying mechanisms. The apparent simplicity of a psychophysical relationship does not mean that the underlying physiological mechanisms are equally simple. Their complexity is indicated by the role of early experience, meaning, and culture on pain perception. These data represent pieces of the puzzle which must play a key role in the development of any satisfactory theory of pain.

Pain, we now believe, refers to a category of complex experiences, not to a single sensation produced by a specific stimulus. We are beginning to recognize the many different qualities of sensory and affective experience that we simply categorize under the broad heading of 'pain'. We are more and more aware of the plasticity and modifiability of events occurring in the central nervous system. Livingston (1943, 1953), long ago, argued against the classical conception that the intensity of pain sensation is always proportional to the stimulus. He proposed instead that pain, like all perceptions, is 'subjective, individual and modified by degrees of attention, emotional states and the conditioning influence of past experience'. Since that time we have moved still further away from the classical assumption that noxious stimulation invariably produces pain, that the pain has only one specific quality, and that it varies only in intensity.

The psychological evidence strongly supports the view of pain as a perceptual experience whose quality and intensity are influenced by the unique past history of the individual, by the meaning he gives to the pain-producing situation and by his 'state of mind' at the moment. We believe that all these factors play a role in determining the actual patterns of nerve impulses that ascend from the body to the brain and travel within the brain itself. In this way pain becomes a function of the whole individual, including his present thoughts and fears as well as his hopes for the future.

3 Clinical Aspects of Pain

The relief of pain and suffering has been a continuing human endeavour since the dawn of recorded history. Yet despite centuries of observation and study, we are only beginning to understand the subtleties and complexities of pain. Even though we now possess many effective 'pain-killing' drugs, we know little about where and how these drugs act. Surgical procedures that are carried out to relieve pain produce dismal failures often enough to convince us that we are far from understanding the neurological mechanisms that subserve pain perception.

The nature of pain is one of the most fascinating problems in medicine and biology. Pain relief, after all, is one of the universal goals of medicine, and an enormous amount of research has been devoted to an understanding of pain mechanisms. At first glance, pain appears to be a simple sensation produced by obviously injurious stimuli, and its relief, presumably, should present few problems. We burn or cut our skin, and we cry out with pain; a stone lodges in our bladder and we feel excruciating pain. In these cases the cause of pain is clear, and normal healing or surgical removal of the diseased organ usually brings about prompt relief.

Other pain problems, however, continue to perplex the clinical investigator. Three pain syndromes, *phantom limb pain*, *causalgia*, and the *neuralgias*, have been studied in detail and present unusual features that are difficult to explain. These types of pain, which begin as signals of serious bodily damage, may persist, spread, and increase in intensity, so that they become maladies in their own right. In each case, the pain may become far worse than that associated with the original injury. In this chapter we will examine the major features of these pain syndromes. They present us with valuable clues about pain mechanisms that we must later consider when we evaluate the current theories of pain.

Phantom limb pain

Phantom limb pain is one of the most terrible and fascinating of all clinical pain syndromes. Its description by Ambroise Paré in 1552 (cited in Keynes, 1952) captures the sense of awe and mystery it evokes in people who hear about it for the first time:

Verily it is a thing wonderous strange and prodigious, and which will scarce be credited, unless by such as have seen with their eyes, and heard with their ears, the patients who have many months after the cutting away of the leg, greviously complained that they yet felt exceeding great pain of that leg so cut off.

Most amputees report feeling a phantom limb almost immediately after amputation of an arm or leg (Simmel, 1956). The phantom limb is usually described as having a tingling feeling and a definite shape that resembles the real limb before amputation. It is reported to move through space in much the same way as the normal limb would move when the person walks, sits down, or stretches out on a bed. At first, the phantom limb feels perfectly normal in size and shape – so much so that the amputee may reach out for objects with the phantom hand, or try to get out of bed by stepping on to the floor with the phantom leg. As time passes, however, the phantom limb begins to change shape. The arm or leg becomes less distinct and may fade away altogether, so that the phantom hand or foot seems to be hanging in mid-air. Sometimes, the limb is slowly 'telescoped' into the stump until only the hand or foot remain at the stump tip.

Although tingling is the dominant sensation of the phantom limb, amputees also report a variety of other sensations, such as pins-and-needles, warmth or coldness, heaviness, and many kinds of pain. About 35 per cent of amputees report pain in the phantom limb at some time (Feinstein, Luce and Langton, 1954). Fortunately, the pain tends to subside and eventually disappear in most of them. In about 5–10 per cent, however, the pain is severe and may become worse over the years. It may be occasional or continuous, and is described as cramping, shooting, burning or crushing. It usually starts immediately after amputation, but sometimes appears weeks, months, even years later. The pain is felt in definite parts of the phantom limb (Livingston, 1943). A common complaint, for example, is that the phantom hand is clenched, fingers bent over the thumb and digging into the palm, so that the whole hand is tired and painful.

If the pain persists for long periods of time, other regions of the body may become sensitized so that merely touching these new 'trigger zones' will evoke spasms of severe pain in the phantom limb (Cronholm, 1951). Pain, moreover, is often triggered by visceral inputs produced by urination and defecation (Henderson and Smyth, 1948). Even emotional upsets such as an argument with a friend may sharply increase the pain. Still worse, the conventional surgical procedures (Figure 4) often fail to bring permanent relief, so that these patients may undergo a series of such operations without any decrease in the severity of the pain. Phenomena such as these defy explanation in terms of our present physiological knowledge.

Before we analyse the major properties of phantom limb pain we shall first examine a case history reported by Livingston (1943, pp. 1–4):

In 1926, a physician, who had long been a close friend of mine, lost his left arm as a result of gas bacillus infection. (He had sustained a puncture wound of his left hand when a glass syringe containing the bacillus broke as he was injecting it into a guinea pig. Virulent and rapidly spreading gas gangrene made its appearance during the night and his arm was amputated the following morning.) The arm was removed by a guillotine type of amputation close to the shoulder and for some three weeks the wound bubbled gas. It was slow in healing and the stump remained cold, clammy, and sensitive . . . . At times the stump would jerk uncontrollably or, after a period of quiet, flip suddenly outward. He suffered a great deal of pain and submitted to a reconstruction operation and the removal of

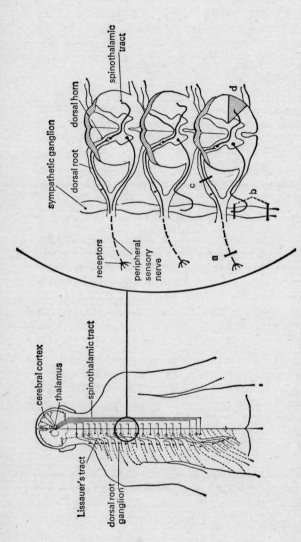

Figure 4 Traditional concept of pain. *Left* is Larsell's (1951) diagram of the pain pathway: pain fibres from each dermatome enter the spinal cord, ascend a few segments (in Lissauer's tract), and connect with fibres that cross the cord and form the spinothalamic tract to the thalamus. Fibres from the thalamus project to the cortex. *Right* is a diagram of spinal cord cross sections and adjacent sympathetic ganglia, showing several neurosurgical procedures to relieve pain. A neurectomy; B sympathectomy; C rhizotomy; and D cordotomy.

neuromas (small nodules of regenerated nerve tissue), without any relief. In spite of my close acquaintance with this man, I was not given a clear-cut impression of his sufferings until a few years after the amputation, because he was reluctant to confide to anyone the sensory experiences he was undergoing. He had the impression, that is so commonly shared by layman and physician alike, that because the arm was gone, any sensations ascribed to it must be imaginary. Most of his complaints were ascribed to his absent hand. It seemed to be in a tight posture with the fingers pressed closely over the thumb and the wrist sharply flexed. By no effort of will could he move any part of the hand . . . . The sense of tenseness in the hand was unbearable at times, especially when the stump was exposed to cold or had been bumped. Not infrequently he had a sensation as if a sharp scalpel was being driven repeatedly, deep into . . . the site of his original puncture wound. Sometimes he had a boring sensation in the bones of the index finger. This sensation seemed to start at the tip of the finger and ascend the extremity to the shoulder, at which time the stump would begin a sudden series of clonic contractions. He was frequently nauseated when the pain was at its height. As the pain gradually faded, the sense of tenseness in the hand eased somewhat, but never in a sufficient degree to permit it to be moved. In the intervals between the sharper attacks of pain, he experienced a persistent burning in the hand. This sensation was not unbearable and at times he could be diverted so as to forget it for short intervals. When it became annoying, a hot towel thrown over his shoulder or a drink of whisky gave him partial relief.

I once asked him why the sense of tenseness in the hand was so frequently emphasized among his complaints. He asked me to clench my fingers over my thumb, flex my wrist, and raise the arm into a hammerlock position and hold it there. He kept me in this position as long I could stand it. At the end of five minutes I was perspiring freely, my hand and arm felt unbearably cramped, and I quit. But you can take your hand down, he said.

He was prepared to submit to a posterior rhizotomy (see Figure 4), but asked my opinion as to whether or not the simpler operation of sympathectomy (Figure 4) might afford some relief. I was unable to predict the effect of a sympathectomy, and suggested that some time when his pain was particularly severe, a novocaine infiltration of the appropriate sympathetic ganglia might provide, by its temporary effect, some index of the value of sympathectomy in his particular case. The opportunity to try this did not occur until early in 1932. On that occasion I was visiting at his home when a particularly bad attack of pain came on. We went at once to the hospital and carried out a novocaine injection of the upper thoracic sympathetic ganglia of both sides. Following the injection the stump was found to be warm and dry, and the pain in the phantom limb gone. To our mutual surprise, he felt that he could voluntarily move each of his phantom fingers. This freedom of movement and complete relief of pain persisted the following day, and when I finished my visit we agreed that the test seemed to indicate that a surgical sympathectomy should be worth doing. It was arranged that he should come to my home city in the next few weeks for this operation.

He did not come, nor did I hear from him for three months. I found then that he had remained entirely free from pain and discomfort in the phantom extremity . . . . He stated that this was the first time he had been free from pain in the phantom hand since the day of amputation. Though he was delighted with the result, he interpreted it as proof of the purely psychic origin of his pains and as confirmation of his fear that he was suffering from a psychoneurosis. He could not see why an injection with novocaine, the effect of which should wear off in a few hours, could possibly confer relief from pain of months' duration. I did not know 'why' it could, but previous experiences with similar injections for other pain syndromes had taught me that it sometimes does confer lasting relief.

The relief from pain persisted for many months but gradually he became aware again of an increasing tension in the phantom hand and an intolerable sensation of constriction in the shoulder, as if 'a wire tourniquet' were being constantly tightened, shutting off the circulation. He had been on a hunting trip in Canada about a month before I saw him in October, 1934. The weather had been chilly and, although he wore a woollen sock over the stump, it had become very cold. He believed that the exposure had aggravated his distress. At this time the stump was extremely cold and wet, measuring 10°–12°C. colder than the same level of the opposite arm. On 12 October 1934, five cc of 2-per-cent solution of novocaine was injected near each of the upper four sympathetic ganglia of the left side. During the placing of the needles, as the second needle was inserted below the neck of the second rib, he complained of a sudden, sharp, stabbing pain in the base of his thumb. The needle was readjusted but the pain persisted. An hour after the injection, all of the digits except the thumb felt warm and relaxed. The thumb seemed to remain pressed into the palm and was the seat of a burning pain. During the night the pain spread up the arm and he slept little in spite of heavy sedation. The following day the burning sensation gradually disappeared. The hand remained warm and the digits were freely movable. For more than seven years the pains did not return. The stump remained warm and insensitive, and there were no attacks of clonic jerking. Sweating remained normal on the affected side. There were intervals in which he seemed to completely forget the phantom arm and at times he could not even voluntarily recall its image. Within recent months, however, there have been signs that trouble might be brewing again. He has had none of his former complaints but occasionally he gets a sharp and arresting twinge of pain in the stump itself. Further treatment may yet be needed in this case.

Properties of phantom limb pain

Phantom limb pain is characterized by four major properties:

1. The pain endures long after healing of the injured tissues. It continues for more than a year after onset in about 70 per cent of patients, and may persist for years, even decades, in patients with perfectly healed stumps (Sunderland, 1968).

2. Trigger zones may spread to healthy areas on the same or opposite side of the body (Cronholm, 1951). Gentle pressure or pin prick on another limb or on the head (Figure 5) may trigger terrible pain in the phantom limb. There is also evidence that pain at a site distant from the stump may evoke pain in the phantom limb. Thus, amputees who develop anginal pain as long as twenty-five years after amputation may suffer severe pain in the phantom limb during each bout of anginal pain, although phantom limb pain may never before have been experienced (Cohen, 1944).

3. Phantom limb pain is more likely to develop in patients who have suffered pain in the limb for some time prior to amputation (Melzack, 1971). It is relatively rare in war amputees who tend to lose a limb suddenly, but more common in civilian amputees, in whom presurgical pain is a frequent accompaniment of pathology of the limb. Furthermore, the pain may resemble, in both quality and location, the pain that was present before amputation (Bailey and Moersch, 1941;

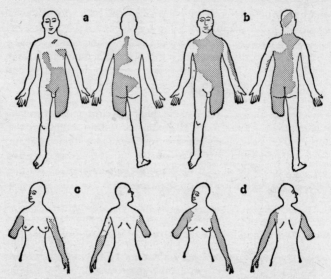

Figure 5 Cronholm's (1951) observations on stimulation sites which evoke pain sensation in the phantom limb. *Top* shows a 59-year-old man who received compound fractures of the lower left leg at the age of twenty-one: amputation was four months later. Pressure (A) or pin pricks (B) were applied to the skin. Stimulation of effective sites (cross hatched areas) produced severe shooting pains and other sensations in the phantom limb. *Bottom* shows a 34-year-old woman: amputation was at the age of fourteen. Pressure (C) or pin pricks (D) were applied to the skin. Stimulation of effective sites (cross hatched areas) produced sensations of a diffuse, unpleasant 'irritation' in the phantom hand.

White and Sweet, 1969). Thus, a patient who was suffering from a wood sliver jammed under a finger nail, and at that time lost his hand in an accident, subsequently reported a painful sliver under the finger nail of his phantom hand. Similarly, lower limb amputees may report pain in particular toes or parts of the phantom foot that were ulcerated or diseased prior to amputation.

4. The pain is sometimes permanently abolished by temporary decreases or increases of somatic input. Injection of local anaesthetics (such as novocaine) into the stump tissues or

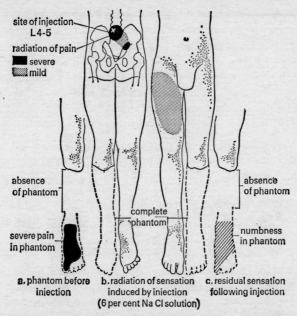

site of injection
L4-5

radiation of pain
■ severe
▨ mild

absence
of phantom

absence
of phantom

complete
phantom

severe pain
in phantom

numbness
in phantom

a. phantom before
injection

b. radiation of sensation
induced by injection
(6 per cent Na Cl solution)

c. residual sensation
following injection

Figure 6 Observations by Feinstein, Luce and Langton (1954)
on the effect of hypertonic saline injection into lumbar (L4–L5)
interspinous tissues on phantom limb pain. The saline injection,
in this case, produced a radiation of pain to the right hip and
thigh, and sudden detailed awareness of the complete phantom
limb. After injection, numbness was felt in the previously painful
area. Pain relief, after this procedure, may last for days, weeks,
sometimes permanently.

nerves may stop the pain for days, weeks, sometimes permanently, even though the anaesthesia wears off within hours
(Livingston, 1943). Successive blocks may produce increasingly longer periods of relief. Similarly, novocaine injected into
the lower-back interspinous tissue in leg amputees produces a
progressive numbness of parts of the phantom limb, and prolonged, sometimes permanent relief of pain in all or part of it
(Feinstein, Luce and Langton, 1954). Paradoxically, increases
in the sensory input may also relieve the pain. Injection of
small amounts of hypertonic saline into the interspinous tissue
of amputees (Figure 6) produces a sharp, localized pain that

radiates into the phantom limb, lasts only about ten minutes, yet may produce dramatic partial or total relief of pain for hours, weeks, sometimes indefinitely (Feinstein, Luce and Langton, 1954). Similar effects have been obtained by hypertonic saline injections into the stump tissues (Nathan, personal communication, 1970). Vigorous vibration of the stump may also produce relief of phantom limb pain (Russell and Spalding, 1950).

These properties define the scope of the problem that confronts us. We shall now examine possible mechanisms to explain them.

The search for causal mechanisms

The mechanisms underlying phantom limb pain have been the basis of bitter controversy. The crux of the problem has been the attempt to discover a single factor as the whole explanation. Historically, the search for *the* causal mechanism has progressed from the periphery to the central nervous system, each site on the way leading to a proposed mechanism and a particular therapy to relieve pain (Figure 4, p. 52). The failure of these therapeutic measures to cure all cases has led to a further explanation – that the patients are neurotic or malingerers and imagining their pain. The evidence, however, is that there is not a single cause. Instead, all of these mechanisms – including emotional disturbance – may contribute to phantom limb pain.

Peripheral mechanisms. Once phantom limb pain is under way, almost any somatic input may augment it. Pressure on tender neuromas or trigger points at the stump can evoke severe, prolonged pain. Despite the obvious contribution of sensory inputs, peripheral irritating factors such as neuromas are clearly not the main cause in most cases (Sunderland, 1968). Surgical section of peripheral nerves or special procedures to prevent neuroma formation frequently fail to stop pain. Even dorsal root section (rhizotomy) is usually ineffective (Sunderland, 1968). A striking example is the case of Henry B., who underwent an extensive rhizotomy. After surgery, he had complete cutaneous anaesthesia from shoulder to umbilicus, but still felt excruciating pain in the phantom fingers which protruded from his upper arm stump (White and Sweet, 1969).

If the cause of the prolonged pain were a chronic irritating lesion, a minor form of therapy such as injection of the affected area, nerve, or roots with anaesthetic drugs could not by itself remove the pathological cause, since the drug effects wear off after two or three hours. The fact that one or more injections may provide prolonged, sometimes permanent, relief rules out irritation at the stump as the major cause (Livingston, 1943). Rather, the data suggest that chronic, low-level sensory inputs *contribute* to the pain, since modulation of the input by anaesthetic blocks clearly influences it, sometimes dramatically.

Sympathetic nervous system mechanisms. The sympathetic nervous system (which affects blood circulation, sweating, and the general nutritive condition of tissues in the limbs) also contributes to the pain in some way (Livingston, 1943; Sunderland, 1968). There are abnormal sympathetic manifestations such as poor blood-flow, coldness, and sweating at the stump. In addition, anaesthetic blocks of the sympathetic ganglia sometimes relieve pain for prolonged periods of time. Sympathetic activity, however, is not the major cause of phantom limb pain. Surgical removal of a portion of the sympathetic ganglia (sympathectomy) rarely produces lasting relief of phantom limb pain (Kallio, 1950). Furthermore, Li and Elvidge (1951) report the case of a paraplegic who sustained a total break in the spinal cord at the level of the chest in addition to a leg amputation, so that the sympathetic ganglia provided the sole afferent route from the stump to the spinal cord. Nevertheless, the pain in his phantom foot was not relieved during a complete bilateral block of the sympathetic ganglia.

Psychological mechanisms. Finally, there is an obvious psychological contribution to phantom limb pain (Melzack, 1971). It is often triggered by emotional disturbances, and is sometimes abolished by distraction-conditioning, hypnosis, and psychotherapy (Merskey and Spear, 1967; Sternbach, 1968). These data, together with the frequent failure of traditional surgical therapy, have led to the suggestion that the patients are in pain because of psychopathological personal needs (Kolb, 1954). It is true that patients suffering phantom limb pain often have emotional disturbances such as anxiety about social adjustment. Indeed the intense, unrelenting pain may itself produce marked withdrawal, paranoia, and other personality changes (Livingston, 1943). However, the hypothesis that phantom limb pain always has a psychiatric basis is untenable. It cannot explain the sudden relief produced by nerve blocks. It would be wrong to assume that the injections have only psychotherapeutic (or placebo) value, because injection of an inappropriate nerve fails to relieve pain, even though injection of the appropriate nerve in the same patient is effective (Livingston, 1943). Moreover, statistical analysis of the data presented by Ewalt, Randall and Morris (1947) indicates that patients with phantom limb pain do not have a greater incidence of neuroses than those without pain in the phantom limb. Emotional factors undoubtedly contribute to the pain but are not the major cause.

In summary, these data, taken together, indicate that phantom limb pain cannot be satisfactorily explained by any single mechanism such as peripheral nerve irritation, abnormal sympathetic activity, or psychopathology. All contribute to the pain in some way. The question is: how? The most satisfactory answer so far is that traumatic or otherwise abnormal inputs may produce a change in information processing in the central nervous system itself, so that abnormal patterns of nerve impulses are triggered by cutaneous and sympathetic inputs and are modulated by brain activities. This concept will be examined in chapters 5 and 6.

Causalgia

Causalgia is a severe, burning pain that is characteristically associated with rapid, violent deformation of nerves by high-velocity missiles such as bullets (Sunderland, 1968). It is estimated to occur in 2–5 per cent of cases of peripheral nerve injury, and is typically seen in young men that have been wounded in military combat. Causalgia persists more than six months after injury in 85 per cent of cases, and then begins to disappear spontaneously. Nevertheless, a year after injury, about 25 per cent still complain of pain (Echlin, Owens and Wells, 1949).

Causalgia (which means 'burning pain') exhibits many of the features of phantom limb pain as well as other unusual characteristics. Its dominant feature is the unrelenting intensity of the pain which evokes images of Dante's *Inferno*. It has been described by patients as being 'like a blaze of fire', 'like someone was pouring boiling water on the top of my foot and holding a cigarette lighter under my big toe', 'like my hand was pressed against a hot stove' (Echlin, Owens and Wells, 1949). Mitchell (1872), who coined the terms 'causalgia' and 'phantom limb pain', believed causalgia to be 'the most terrible of all tortures which a nerve wound may inflict'.

The classic description of causalgia was recorded by Mitchell (1872, pp. 292–6) at the time of the American Civil War. He describes the case of Joseph Corliss, who was shot in the left arm by a bullet that entered just above the elbow, penetrated without touching the artery, and emerged through the belly of the biceps:

On the second day the pain began. It was burning and darting. He states that at this time sensation was lost or lessened in the limb, and that paralysis of motion came on in the hand and forearm. The pain was so severe that a touch anywhere, or shaking the bed, or a heavy step, caused it to increase.

In an attempt to stop the pain, the wounded area was opened surgically, a small portion of the damaged median nerve was excised, and the cut ends were sutured together:

The man states, very positively, that the pain in the median distribution did not cease, nor perceptibly lessen, but that he became more sensitive, so that even the rattling of a paper caused extreme suffering. He 'thinks he was not himself' for a day or two after the operation. It seems quite certain that the pain afterwards gradually moderated .... Meanwhile the hand lay over his chest, and the fingers, flexing, became stiffened in this position.

The pain persisted despite healing of the wound, so that two years after the injury:

He keeps his hand wrapped in a rag, wetted with cold water, and covered with oiled silk, and even tucks the rag carefully under the flexed finger tips. Moisture is more essential than cold. Friction outside of the clothes, at any point of the entire surface, 'shoots' into the hand, increasing the burning (pain) .... Deep pressure on the muscles has a like effect, and he will allow no one to touch his skin, save with a wetted hand, and even then is careful to exact careful manipulation. He keeps a bottle of water about him, and carries a sponge in the right hand. This hand he wets before he handles anything; used dry, it hurts the other limb. At one time, when the suffering was severe, he poured water into his boots, he says, to lessen the pain which dry touch of friction causes in the injured hand .... He thus describes the pain at its height: 'It is as if a rough bar of iron were thrust to and fro through the knuckles, a red-hot iron placed at the junction of the palm and (thumb), with a heavy weight on it, and the skin was being rasped off my finger ends.'

The debilitating effects of such prolonged pain have been described by Mitchell (pp. 196–7):

Perhaps few persons who are not physicians can realize the influence which long-continued and unendurable pain may have upon both body and mind The older books are full of cases in which, after lancet wounds, the most terrible pain and local spasms resulted. When these had lasted for days or weeks, the whole surface became hyperaesthetic, and the senses grew to be only avenues for fresh and increasing tortures, until every vibration, every change of light, and even ... the effort to read brought on new agony. Under such torments the temper changes, the most amiable grow irritable, the soldier becomes a coward, and the strongest man is scarcely less nervous than the most hysterical girl.

Role of input from the limb

An abnormal sensory input from the areas innervated by the injured nerve is clearly implicated in causalgia (Livingston, 1943). The fact that even the gentlest touch may provoke pain makes these people withdraw from all tactile stimuli. They tend to protect the limb by placing wet clothes around it, and keep it almost immobile since any movement is usually accompanied by pain. The pain, then, limits movement, which in turn decreases the usual, patterned cutaneous and proprioceptive input from the limb. The input, therefore, is doubly abnormal as a result of the lesion of the nerve and the excessive protection of the limb. Any momentary freedom from pain is, of course, accompanied by overall increases in input through cutaneous and proprioceptive channels.

Once the causalgic state is full-blown, the original injury is no longer the major cause of pain. The frequent failure of peripheral nerve surgery to abolish pain indicates that more is involved than simply an irritating peripheral lesion. Section of the peripheral nerve at successively higher levels, amputation of the limb, and cutting the dorsal sensory roots which enter the spinal cord, have all produced as many failures as successes. Indeed, operations have been performed for causalgic pain at nearly every site in the sensory pathway from peripheral receptors to somatosensory cortex, and at every level the story is the same: some initial encouraging results, but a disheartening tendency for the pain to return (Sunderland, 1968).

Modulation of the sensory input, however, may bring about dramatic relief of pain. Injection of local anaesthetics into the nerves or tissues associated with the lesion may abolish pain for hours or days, and on rare occasions it never returns. Livingston (1948) reports, moreover, that the pain can be abolished if the patient is trained to tolerate sensory stimulation of the affected limb and is urged to use it normally. He encouraged his patients to place the affected arm in warm water baths, in which the water currents flowing over the limb were made increasingly vigorous over a period of weeks. Once the patient allowed the therapist's hand to touch the limb under water, he was urged to permit the therapist to massage the arm, at first gently, then more vigorously. As the pain diminished, the patient tended to use the arm, which in turn produced a more normal proprioceptive input. In this way, the causalgic pain diminished in intensity over a period of weeks.

Non-specific triggering stimuli

One of the most remarkable features of causalgic pain is the degree to which it is triggered or enhanced by a variety of non-noxious stimuli. Pain is produced by the gentlest somatic stimulation, and even by non-somatic stimuli. Sudden noises, rapidly changing visual stimuli, emotional disturbances, almost any stimulus which elicits a startle response, are all capable of making the pain worse (Livingston, 1943). When Livingston was Commander of the Peripheral Nerve Injury Ward in a United States Naval Hospital, he had to request a special order from Washington to prevent airplanes from flying in the vicinity of the hospital since the vibration and noise produced screams of agony in his causalgic patients.

Sympathetic nervous system mechanisms

The sympathetic nervous system appears to play a particularly important role in causalgia. The affected limb usually shows a variety of symptoms indicative of abnormal sympathetic activity. The hand is cold, often drips sweat, is discoloured (presumably due to vascular changes) and even the finger-nails become brittle and shiny. Injection of a local anaesthetic (such as novocaine) into the sympathetic ganglia may dramatically abolish the pain as well as the abnormal sympathetic symptoms for long periods of time, sometimes permanently (Livingston, 1943). Sympathectomy, moreover, usually produces permanent relief of causalgia (Sunderland, 1968).

These observations have led some writers to assume that 'pain fibres' travel through the sympathetic ganglia. Sympathectomy, however, is much more effective as a cure for causalgia than it is for phantom limb pain (Kallio, 1950). It is difficult to conceive of sympathetic 'pain fibres' that specifically evoke causalgia but not phantom limb pain. Sympathetic fibres have been invoked in yet another role. It has been proposed, to explain causalgia, that sympathetic efferent fibres directly activate somatic afferent fibres at the site of the lesion by way of pathological fibre-to-fibre connections called ephapses (Barnes, 1953). This explanation, however, fails to account for the relief of pain afforded by blocks of the somatic nerve *distal* to the lesion – that is, between the nerve lesion and the skin (Livingston, 1943; see Kibler and Nathan, 1960). Moreover, rhizotomy should prevent ephaptic connections from transmitting signals into the spinal cord, yet the pain often persists after the operation.

In summary, the data show that causalgia, like phantom limb pain, is determined by several contributions: by sensory inputs from the somatic as well as the auditory and visual systems, by activity in the sympathetic nervous system, and by cognitive activities such as emotional disturbance. No one of these contributions can be considered as the sole cause. Rather, the data suggest that causalgia is brought about by changes in activity in the central nervous system, so that all avenues of input are now capable of triggering nerve impulse patterns that produce pain. Abnormal sympathetic manifestations are clearly a major source of somatic input, and appear to have a more potent role in causalgia than in phantom limb pain. Nevertheless, they represent only one source of sensory input. Cutaneous and proprioceptive inputs also evoke pain, and they too can be modulated to bring about temporary or permanent relief of pain.

The neuralgias

There are several pain syndromes associated with peripheral nerve damage that are generally categorized as neuralgic pain. Their properties are essentially similar to those of phantom limb pain and causalgia, and are characterized by severe, unremitting pain which is difficult to treat by surgical or other traditional methods. The causes of neuralgic pain include viral infections of nerves, nerve degeneration associated with diabetes, poor circulation in the limbs, vitamin deficiencies, and ingestion of poisonous substances such as arsenic or lead. In brief, almost any infection or disease that produces damage to peripheral nerves, particularly the large myelinated nerve fibres, may be the cause of pain that is labelled as neuralgic.

Post-herpetic neuralgia

Infection by the virus *herpes zoster* (which is related to the virus that causes chicken pox) produces inflammation of one or more sensory nerves. The inflammation, which is painful, is associated with eruptions (or 'shingles') at the skin along the course of the nerve. The herpetic attack is itself painful, but the pain usually subsides as new nerve fibres regenerate into the skin. In a small number of people, however, the post-herpetic pain persists and may become worse. Noordenbos (1959) notes that neuralgic skin areas are not only the site of spontaneous pain (in the absence of stimulation), but are extremely hyperaesthetic, so that the pain is aggravated by any cutaneous stimuli applied to them. Even the friction of clothes is highly unpleasant and contact is avoided as much as possible. The pain may also be intensified by noise in the immediate vicinity or by emotional stress. This condition may last for many months or even years, and is extremely resistant to most conventional forms of therapy including surgical treatment.

Noordenbos (1959) describes two major characteristics of post-herpetic pain. The first is the remarkable summation of stimulation. One of the stimuli Noordenbos used was a test-tube containing hot water. When the hot tube was applied to normal skin, the patient reported that it felt hot but tolerated it without discomfort for long periods of time. When it was then placed on the neuralgic skin area, an entirely different sequence of events occurred. There was no sensation of temperature for the first few seconds. The tube was then gradually felt as warm or tingling, slowly becoming hotter. If stimulation was continued, the patient stated that it began to burn and finally he cried out with pain and pushed the examiner's hand away. This whole sequence took from twenty seconds to as long as a full minute or longer. Noordenbos notes that if a larger surface of the hot tube was applied to the skin the entire sequence was accelerated, starting as indifferent and rapidly going through all the intermediate sensations to end in unbearable pain. Thus the speed of summation of input was dependent on the size of the area that was stimulated.

The second characteristic is a marked delay in the onset of pain after stimulation. This was apparent in the sequence of events Noordenbos observed after application of the hot test-tube. It was especially clear when he applied multiple gentle pin pricks to the affected skin areas. After a distinct delay following onset of the pin pricks, the patients reported feeling intense pain that spread over large areas and then wore off slowly. The onset of pain was 'very sudden, almost explosive in character, and had an extremely unpleasant quality that differed markedly from the pain evoked in normal skin with the same stimulus' (Noordenbos, 1959, p. 8).

Trigeminal neuralgia

Several neuralgic states are associated with the nerves of the head and face, and are classified on the basis of the particular nerve which is affected (White and Sweet, 1969). Trigeminal neuralgia – which is also known as 'tic douloureux' – is particularly vicious. It is characterized by paroxysmal attacks of pain that may be triggered by eating or talking, or may even occur spontaneously. The pain is extremely severe, so that these people often refuse to eat or talk, and become physically weak and depressed. These properties are described in a case history reported by Livingston (1943, pp. 147–8):

Mr J.M., aged sixty-four, suffered from attacks of trigeminal neuralgia in 1932. In 1933 a competent neurosurgeon partially divided the posterior root of the trigeminal nerve . . . . He was completely relieved of pain for three years. Then, in spite of the fact that there was numbness . . . the attacks, exactly similar in type, recurred. A second operation was carried out to divide the posterior root more completely. Again he had a period of three years of complete relief. In 1939 the pain attacks again began, in all respects similar to his previous trouble. He was advised to submit to a third operation . . . to divide the posterior root closer to its brain connection. He refused the third operation. At the time of my first examination in May 1940, he said that his pain was beyond description, in spite of large doses of hypnotics and opiates. Eating, smoking, talking, shaving, or brushing his teeth brought on the painful paroxysms.

He had lost forty pounds in weight .... Attacks could be set off by contact near the ... nostril, near the outer margin of the lip, near the ... eye, and at two areas in the gum on each side of his two remaining incisor teeth in the right upper jaw. Each of these sensitive points was injected several times in the course of a three weeks' treatment. Whenever he had a twinge of pain, the site from which the pain seemed to originate was infiltrated with novocaine. The attacks rapidly diminished in frequency and severity. There was a temporary exacerbation when the two incisor teeth were removed. He remained entirely free from pain from June until December. Then he had a mild recurrence requiring five injections to control. In March 1942, sharp paroxysms began again. They were relieved within two weeks by a few injections and the fitting of an upper plate.

One of the remarkable features of tic douloureux is that the paroxysmal attacks are triggered by gentle stimulation but not by intense stimuli (Kugelberg and Lindblom, 1959; White and Sweet, 1969). Severe pinches, pin jabs, or intense pressure, heat or cold applied to the trigger zones usually fail to evoke pain. When weak stimuli are used, however, paroxysms are evoked after a long summation time. Successive gentle touches for fifteen to thirty seconds may be required to fire an attack. The pain may then last for one to three minutes. These observations, together with the fact that there is a refractory period of several minutes after an attack before a new one can be evoked, led Kugelberg and Lindblom (1959) to conclude that tic douloureux is the result of abnormal central neural processes. Fortunately, tic douloureux is one of the few pain states which can often be treated simply and effectively. Tegretol (carbamazepine), a drug which is used to treat epilepsy, also relieves tic douloureux in the majority of patients (White and Sweet, 1969). Unfortunately, it appears not to be effective in the treatment of other neuralgic pain states.

In summary, the neuralgias exhibit many of the properties characteristic of phantom limb pain and causalgia. The remarkable summation of cutaneous stimulation at hyperaesthetic skin areas, the increases in pain produced by sudden auditory stimulation or by emotional disturbance, and the long delays in perception all point to abnormal activity in the central nervous system. Particularly valuable clues on the nature of pain have been obtained from studies of post-herpetic and trigeminal neuralgia. Noordenbos (1959) and Kerr and Miller (1966) have found that both pain states are associated with a selective loss of the large myelinated fibres and a proliferation of the small unmyelinated fibres. The fact that these syndromes are usually found in older people coincides with the fact that the myelinated peripheral nerve fibres degenerate during ageing, so that about 30 per cent of the fibres are lost by age sixty-five (Gardner, 1940). The importance of this observation will become apparent in later chapters on pain theories. These changes in peripheral nerves, however, cannot be the whole story: surgical section of the appropriate nerves right up to the point of entry into the spinal cord or brain frequently fails to relieve the pain. Rather, the data suggest that changes in central nervous system activity, perhaps initiated by peripheral factors, may underlie the summation, delays, persistence and spread of pain.

Post-traumatic pain syndromes

The final syndrome we will consider has been labelled 'post-traumatic pain'. It may occur after various kinds of accidents, and the ensuing pain usually persists long after healing is presumed to have occurred. The severity of pain, moreover, often exceeds that which is expected from the minor bruises or lacerations which mark the beginning of some of these syndromes. If the syndromes are not brought under control, the pain and trigger zones may spread to other areas of the body. Pain originally limited to the cheek, for example, may spread to the lower jaw and forehead. Fortunately, some of these cases are cured by successive anaesthetic blocks (Livingston, 1943, pp. 118–19):

Mr M.G., a 32-year-old truck driver struck the ulnar side of his right forearm against the end-gate of his truck. X-rays did not reveal a fracture and the local discolouration and swelling did not last long. But he complained of constant pain which was not relieved by physiotherapy treatments carried out for many months. He complained of weakness of the grip of the right hand and a tendency to drop things. He stated that when he drank a cup of coffee he had to hold the left hand under the cup for fear he would unexpectedly drop it. The whole arm was said to ache, and the hand was cold and often of a dusky colour. He said the forearm and hand felt 'half asleep and numb'.

Examining physicians had told the insurance carriers that they found no objective evidence of anything wrong with the hand and arm except for a slight coldness. The officials of the insurance company were convinced that he 'didn't want to return to work', but his family physician, who had known the man for years, was equally insistent that 'something was still wrong'. Examination showed that there was an accurately localized spot of tenderness remaining at the site of his original injury, and when a few cubic centimetres of 2 per cent novocaine solution had been infiltrated into this area there was an immediate improvement in the strength of his grip and a disappearance of his pain. The arm was 'sore' for a few days after the injection but the pain did not return and within a week he was back at work.

Cases such as these are among the most difficult to diagnose and treat. The severity and persistence of pain after accidents from which most people recover uneventfully often raise the possibility of predominantly psychological causes. This may be true in some cases. Yet in others, such as the one just described, the fact that effective relief eventually occurred seems to preclude the possibility that the patient was simply malingering. Indeed, the inability to help these patients may be more revealing of our ignorance about pain mechanisms than of the patients' personalities. The cause of pain is characteristically elusive in these cases. For example, the site of successive intravenous injections later become a focus of burning, persistent pain in one of Livingston's (1943) patients. It is cases such as these that suggest the possibility that prolonged, low-level inputs may bring about long-term changes in activity in the central nervous system, so that pain may persist and spread long after healing has occurred.

Implications of the clinical evidence

The implications of the pathological pain syndromes described above are the following:

1. *Summation.* Gentle touch, warmth, and other non-noxious somatic stimuli can trigger excruciating pain. The fact that repetitive or prolonged stimulation is usually necessary to elicit pain, together with the fact that referred pain can often be triggered by mild stimulation of normal skin, makes it unlikely that the pain can be explained by postulating hypersensitive 'pain receptors'. A more reasonable explanation is that abnormal information processing in the central nervous system allows these remarkable summation phenomena to occur.

2. *Multiple contributions.* The pain, in these syndromes, cannot be attributed to any single cause. There are, instead, multiple contributions. The cutaneous input from the affected part of the body obviously plays an important role. However, inputs that result from sympathetic activity are also important. So too are inputs from the auditory and visual systems. All of these inputs appear to act on structures in the central nervous system that summate the total activity to produce nerve impulse patterns that ultimately give rise to pain. Anxiety, emotional disturbance, anticipation and other cognitive activities of the brain also contribute to the neural processes underlying these pains. They may facilitate or inhibit the afferent input and thereby modulate the quality and severity of perceived pain.

3. *Delays.* Pain from hyperalgesic skin areas often occurs after long delays and continues long after removal of the stimulus. Gentle rubbing, repeated pin pricks, or the application of a warm test-tube may produce sudden, severe pain after delays as long as forty-five seconds. Such delays cannot be attributed simply to conduction in slowly conducting fibres; rather, they imply a remarkable temporal and spatial summation of inputs in the production of these pain states.

4. *Persistence.* The durations of these pain states often exceed the time taken for tissues to heal or for injured nerve fibres to regenerate. Causalgia tends to disappear as regeneration occurs, but sometimes it persists for years, as does neuralgic or phantom limb pain. Furthermore, in all of these syndromes pain may occur spontaneously for long periods without any apparent stimulus. These considerations – together with the observation that pain in the phantom limb frequently occurs at the same site as it occurred in the diseased limb prior to amputation – suggest the possibility of a memory-like mechanism in pain.

5. *Spread.* The pains and trigger zones may spread unpredictably to unrelated parts of the body where no pathology exists. This is further evidence that the central neural mechanisms involved in pain receive inputs from multiple sources. The organization of these mechanisms does not reflect the precise dermatomal (or segmental) innervation of the body by the somatic nerves. (This is immediately evident when Figures 4 and 5 on pp. 52 and 56 are compared.) Instead, the mechanisms appear to be more widespread and receive inputs from all parts of the body.

6. *Resistance to surgical control.* The widespread distribution of the neural mechanisms associated with these pain states is also indicated by the frequent failure to abolish pain by surgical methods. Surgical lesions of the peripheral and central nervous systems have been singularly unsuccessful in abolishing these pains permanently, although the lesions have been made at almost every level from receptors to sensory cortex. Even after such operations, pain can often still be elicited by stimulation below the level of section and may be more severe than before the operation.

7. *Relief by modulation of the sensory input.* The most promising method of treatment for these pains appears to be the modulation of the sensory input by either decreasing or increasing it. Phantom limb pain is sometimes relieved by successive local anaesthetic blocks of tender areas, peripheral nerves or sympathetic ganglia. It may also be relieved by

vigorous vibration or by pain-producing injections of hypertonic saline into the stump or low-back interspinous tissues. Causalgia and the neuralgias can similarly be helped by anaesthetic blocks that temporarily decrease inputs from the affected areas, or by increased stimulation such as vigorous massage. These observations have given rise to new therapeutic methods that hold great promise for the relief of pain without producing irrevocable damage to the peripheral or central nervous systems.

These properties and their implications provide valuable clues towards an understanding of pain. They represent parts of a puzzle which, together with those obtained from psychology and physiology, will reveal the solution to a perplexing, urgent problem. Any satisfactory theory of pain must be able to explain the properties of these syndromes. If our theories do not lead eventually to effective treatment, they have failed, no matter how elegant or compelling they may seem. The clinical problems of pain, in other words, represent the ultimate test of our knowledge.

4 The Physiology of Pain

The psychological and clinical phenomena of pain provide a framework for the physiological problems we will consider in this chapter. Pain, as we have seen, is a highly personal, variable experience which is influenced by cultural learning, the meaning of the situation, attention, and other cognitive activities. How does the central nervous system function to permit such powerful cognitive control over the somatic sensory input? Pain, it is generally acknowledged, is primarily a signal that body tissues have been injured; yet pain may persist for years after tissues have healed and damaged nerves have regenerated. How can we account for neurophysiological processes that go on for such long durations? Similarly, pain and trigger zones sometimes spread to distant, unrelated parts of the body. Can we understand such phenomena in terms of the known connections among neurons in the nervous system? While present-day physiology has answers to some of these problems, it is not even close to explaining others. The physiological and anatomical data are no simpler than the psychological and clinical phenomena of pain.

Two terms are especially critical in our attempts to understand the physiology of pain: *specificity* and *specialization*. *Specificity* implies that a receptor, fibre, or other component of a sensory system subserves only a single specific modality (or quality) of experience. *Specialization* implies that receptors, fibres, or other components of a sensory system are highly specialized so that particular types and ranges of physical energy evoke characteristic patterns of neural signals, and that these patterns can be modulated by other sensory inputs or by cognitive processes to produce more than one quality of experience or even none at all. It is the latter approach – specialization of function – that provides the conceptual framework for this chapter.

It is customary to describe the somatosensory system by proceeding from the peripheral receptors to the transmission routes that carry nerve impulses to areas in the brain. However, it is essential to remember that stimulation of receptors does not mark the beginning of the pain process. Rather, stimulation produces neural signals that enter an active nervous system that (in the adult organism) is already the substrate of past experience, culture, anticipation, anxiety and so forth. These brain processes actively participate in the selection, abstraction and synthesis of information from the total sensory input. Because sensory physiological processes are complex, a brief outline of the somatic sensory system will be provided first, and each step will then be examined in more detail. Only later, when we analyse the contemporary theories of pain, will we try to choose the data that seem most relevant and put them together in a way that is consistent with the psychological and clinical data.

Outline of somatic sensory mechanisms

We may ask at this point: what is the nature of the sensory nerve signals or messages that travel to the brain after injury? Let us say a person has burned a finger; what is the sequence of events that follows in the nervous system? To begin with, the intense heat energy is converted into a code of electrical nerve impulses. These energy conversions occur in nerve endings in the skin called receptors, of which there are many different types. It was once popular to identify one of these types as the specific 'pain receptors'. We now believe that receptor mechanisms are more complicated. There is general agreement that the receptors which respond to noxious stimulation are widely branching, bushy networks of fibres that penetrate the layers of the skin in such a way that their receptive fields overlap extensively with one another (Figure 7). Thus damage at any point on the skin will activate at least two or more of these networks and initiate the transmission of trains of nerve impulses along sensory nerve fibres that run from the finger into the spinal cord. What enters the spinal cord of the central nervous system is a coded pattern of nerve impulses, travelling along many fibres and moving at different speeds and with different frequencies.

Before the nerve-impulse pattern can begin its ascent to the brain, a portion of it must first pass through a region of short, densely packed nerve fibres that are diffusely interconnected. This region, found throughout the length of the spinal cord on each side, is called the substantia gelatinosa (Figure 7). It is in the course of transmission from the sensory fibres to the ascending spinal cord neurons that the pattern may be modified.

Once the sensory pattern has been transmitted to the spinal cord neurons, it projects to the brain along nerve fibres that occupy the anterolateral (front and side) portions of the spinal cord. Many of these fibres continue to the thalamus, forming the spinothalamic tract. The majority of the fibres, however, penetrate a tangled thicket of short, diffusely interconnected nerve fibres that form the central core of the lower part of the brain (Figure 7). This part of the brain is called the reticular formation, and it contains several highly specialized systems which play a key role in pain processes. From the reticular formation, there emerges a series of pathways, so that sensory patterns now stream along multiple routes to other regions of the brain.

Factual information about the afferent processes related to pain really ends at this point. We know that nerve impulses are projected to the cortex but, because large cortical lesions rarely diminish or abolish pain, it is assumed that the cortical projections represent only one of several pathways involved in pain. Other pathways project to the limbic system that forms an extensive and important part of the brain. Moreover, there is evidence that the cortex is not a final destination (or 'pain centre') but that it processes the information it receives and transmits it to deeper portions of the brain. In short, the afferent process from skin to cortex marks only the beginning of prolonged, interacting activities.

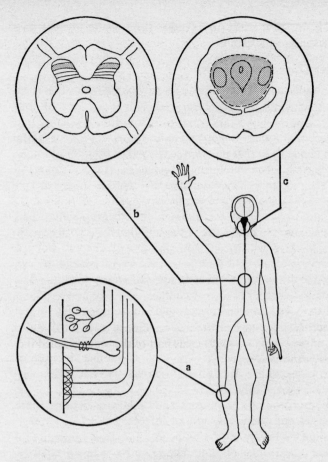

Figure 7 Schematic representation of the receptors and projection pathways of the somatic sensory system. A: The diagram of the skin shows widely branching free nerve-endings (which produce overlapping receptive fields) as well as some specialized end-organs. The fibres project to the spinal cord. B: The cross section of the spinal cord shows the laminae (layers) of cells in the dorsal horns which receive sensory fibres and project their axons toward the brain. The cross-hatched area represents the substantia gelatinosa (laminae 2 and 3). C: The brainstem (lower part of the brain) receives a large somatosensory input, and projects to higher as well as lower areas of the central nervous system. The cross-hatched area represents the reticular formation. Below it on each side is the medial lemniscus. The spinothalamic projections – which are shown within the reticular formation – lie above the lemniscal tracts.

We are now ready to examine the somatic afferent processes in greater detail.

Receptor mechanisms

The traditional picture of skin sensitivity maintains that there are four kinds of receptors, each subserving one of four modalities of cutaneous sensation: pain, touch, warmth and cold. Each receptor is assumed to have a sensitive 'spot' at the skin above it, so that pain sensitivity at the skin, for example, is believed to take the form of discrete 'pain spots' subserved by 'pain receptors'. According to this concept, the free nerve-endings are specific pain receptors while the more complex receptor organs (Figure 7) subserve the other modalities. This simple concept is the basis of the traditional *specificity theory of pain* which will be discussed in the next chapter. For the present, it is sufficient to note that the free nerve-endings can give rise to the full range of cutaneous sensory qualities. The pinna (outer part) of the ear contains only free nerve-endings and the specialized endings around hair follicles. Yet we feel warmth, cold, touch, itch, tickle, pain, or erotic sensations when these areas are appropriately stimulated (Sinclair, 1967).

Histologists (anatomists who look at the fine structure of body tissues) have discovered a rich variety of receptor endings in skin and other tissues. The most common of all are the free nerve-endings. Figure 7 shows how a sensory fibre branches out extensively so that its receptive field – the skin area innervated by all the branches of a single nerve fibre – covers a wide area of skin. The receptive fields of adjacent fibres overlap one another, so that stimulation of a spot of skin activates not one but several fields (Tower, 1943).

Receptor specialization

There are many recent studies in which physiologists have recorded from a single fibre, applied various kinds of stimuli to its receptive field, and thereby determined its properties. These studies have found an astonishing degree of specialization among receptor-fibre units. Receptive fields vary in size and shape, some are highly sensitive to narrow ranges of one or two types of stimuli (such as pressure and temperature), they have different adaptation rates, and so on. Zotterman (1959) has studied temperature-sensitive receptor-fibre units in the cat's tongue, and found that each unit is sensitive to a narrow range of temperatures and shows a peak firing frequency to an even narrower range. There are not simply cold fibres or warmth fibres, but rather each fibre is associated with a receptor that is finely tuned to produce a peak firing rate when a particular temperature is applied. An important point that emerges from Zotterman's studies is that receptor-fibre units are continuously (or 'tonically') active so that their firing rates may increase or decrease depending on the direction of the temperature change.

There is now considerable evidence that many skin receptors respond to at least two classes of environmental energy (Hunt and McIntyre, 1960; Bessou and Perl, 1969). A large number of receptor-fibre units are sensitive to both pressure and temperature, and even units that respond to hair movement may also respond to temperature change. This does not mean, however, that these receptors respond to the full range of all environmental stimuli. Rather, each responds within a narrow thermal range and has a distinct threshold to pressure. The physiological properties that determine receptor specialization thus appear to be highly complex. Melzack and Wall (1962, p. 343) have noted that:

The transduction properties of any given receptor are a function of at least eight physiological variables: (1) threshold to mechanical distortion, (2) threshold to negative and positive temperature change, (3) peak sensitivity to temperature change, (4) threshold to chemical change, (5) stimulus strength–response curve, (6) rate of adaptation to stimulation, (7) size of receptive field, and (8) duration of after-discharge. It is suspected that many of these variables are interrelated.

Since there is reason to believe ... that each of these variables has a continuous distribution, the specialization of any given receptor can be specified accurately in terms of its coordinates with respect to a number of the variables. Thus it is possible to define a particular receptor-fibre unit by saying, for example, that it has a low pressure threshold, a peak sensitivity at high temperature, a narrow receptive field, and a fast rate of adaptation. The specialization of each skin receptor would therefore be defined in terms of its position in a multidimensional space of physiological variables. If receptors are distributed throughout the space defined by these variables, the degree of specialization and number of different kinds of receptors must be very great indeed.

Melzack and Wall (1962) have proposed, moreover, that a receptor generates temporal patterns of nerve impulses rather than modality-specific impulses. These patterns would be determined by the effects of the physical stimulus within the limits set by the receptor's physiological properties such as its sensitivity range and rate of adaptation. Since many receptors respond to at least two different kinds of energy, they must be capable of generating more than one kind of temporal pattern. And, indeed, receptor-fibre units have been observed to respond to tactile stimuli with a characteristically higher firing frequency than is produced by thermal stimulation (Hunt and McIntyre, 1960).

Skin sensitivity and receptive fields

Recent studies of skin sensitivity provide a picture that is more consistent with the evidence of overlapping receptive fields than with the concept of a mosaic of sensitive spots. Maps of thermal sensitivity of large areas of skin (Figure 8) show highly sensitive areas surrounded by regions of decreasing sensitivity (Melzack, Rose and McGinty, 1962). These large sensitive areas, observed by psychological mapping techniques, undoubtedly reflect the activity of overlapping receptive fields that project to successive levels in the central nervous system.

Skin spots, then, appear to represent areas of peak sensi-

Figure 8 Distribution of cold sensitivity of approximately a quarter of the back. The position and size of the area tested is drawn in the lower right corner. The sensitivity distribution was mapped with a round stimulator tip 2·5 mm in diameter, at a temperature of 10°C. The cold intensities reported by the subject are represented by different shades of stipple.
(from Melzack, Rose and McGinty, 1962, p. 300)

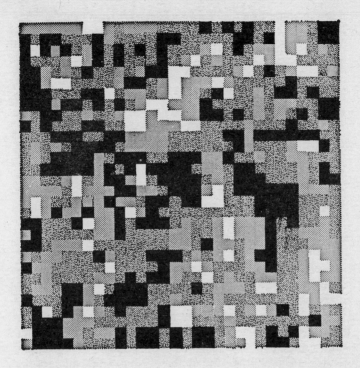

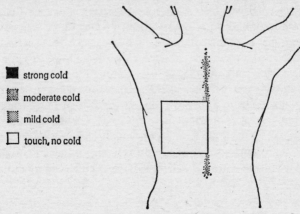

■ strong cold

▨ moderate cold

▧ mild cold

□ touch, no cold

tivity surrounded by valleys of lesser sensitivity. The fact that receptive fields overlap extensively makes it almost certain that even a pin-prick will activate several receptor-fibre units. Thus, the evidence suggests that the 'skin spot', once believed to represent a single specific receptor lying beneath it, is the result of the ability of the central nervous system to integrate the impulses of many fibres having extensively overlapping receptive fields (Tower, 1943).

Even more remarkable are the observations of continuous shifts in skin sensitivity (Melzack, Rose and McGinty, 1962). Successive maps show that large sensitive fields may 'fragment' and 'coalesce', and thereby produce continually changing patterns of sensitivity distribution. It seems likely that these fluctuations represent changes in information transmission throughout the somatosensory projection system. They may be due to changes in activity in receptor-fibre units as well as at synapses throughout the transmission system. Changes in the receptive fields of central cells have been observed by Nakahama, Nishioka and Otsuka (1966). After they mapped the receptive fields of single cells in the thalamus, they injected anaesthetic solutions into the skin so that the cells could no longer be driven by stimulation within the fields. However, they found that within a few minutes the cells were driven by stimulation of adjacent skin areas that had previously had no effect on them (Figure 9). It is apparent, then, that a central cell normally has a large skin area that can drive it (the receptive fields of many fibres that project on to the central cell), but that only a portion of the fibres is capable of doing so at any time. Because receptive fields may vary in size and sensitivity from moment to moment, any transient input – such as a stimulus – produces activity that must be selected by the brain from a continually changing background.

The receptive fields of central cells may also be modulated by brain activities. Taub (1964) found that stimulation of portions of the brainstem activated descending fibres that reduced the size of the receptive fields of cells in the spinal cord. It is noteworthy that lesions in these same areas in cats produced behavioural evidence of hyperalgesia (Melzack,

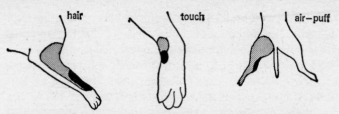

Figure 9 Appearance of new receptive fields. Receptive fields (indicated in black) of cells in the thalamus were first mapped by using three kinds of stimuli: hair movement, touch, and air puffs. After the receptive fields were anaesthetized by localized subcutaneous injections of novocaine, mapping procedures were again carried out, demonstrating the emergence of new receptive fields (stippled areas).
(from Nakahama, Nishioka and Otsuka, 1966, p. 180)

Stotler and Livingston, 1958). Many of the cats were excessively responsive to noxious stimuli and some of them showed behaviour suggesting spontaneous pain in the absence of external stimulation. It is possible that the lesion produced a release from inhibition and an increase in the size of the receptive fields so that noxious stimulation now activated a larger-than-normal number of receptor-fibre units. Consequently, stimuli that tended to be mildly painful now produced much more intense pain.

Receptors in the skin, then, cannot be considered in isolation but can only be understood in terms of their relation to adjacent receptors and their projections to successive transmission levels in the central nervous system. Their sensitivity fluctuates, possibly because of blood flow or other changes in the skin, and the information they project through the central nervous system is also modified by messages descending from the brain. The brain receives impulses projected from many receptors, and these must be synthesized to produce the neural activities that are eventually felt as pain, touch, or any of the other cutaneous sensations.

Somatic sensory nerves

The nerve impulses generated by receptors in body tissue are transmitted along nerve fibres to several destinations in the spinal cord and brain. Each somatic nerve contains fibres of different sizes, and it is now well established that the larger the fibre, the faster it conducts nerve impulses. Generally speaking, there are two kinds of fibres: myelinated and unmyelinated. In peripheral nerves, unmyelinated fibres are three or four times as numerous as myelinated fibres (Sinclair, 1967). The myelinated fibres are also known as A fibres, the unmyelinated as C fibres. Their conduction rates range from about 120 metres per second (for the largest A fibres) to about 1 metre per second (for the smallest C fibres). Moreover, the A fibres consist of distinct subgroups that have been labelled alpha, beta, gamma and delta. The reason for this is the observation that the action potential – the composite electrographic picture of all nerve impulses transmitted along a nerve bundle when one end is electrically stimulated – shows the A wave to have distinct bumps, which represent groups of neurons with particular conduction rates.

The discovery of these distinct fibre groups led to ingenious experiments (reviewed by Sinclair, 1967) in the attempt to associate each fibre group with a specific cutaneous modality and, by inference, with a particular type of receptor. Pain, for example, is assumed by many investigators to be due specifically to activation of receptors with high thresholds to pressure or temperature ('nociceptors'), and attempts have been made to discover the 'pain fibres' attached to the nociceptors. In the search for peripheral fibres that respond exclusively to high-intensity stimulation, Burgess and Perl (1967) have recently discovered a specialized class of A-delta fibres that transmit impulses only when the skin is actually damaged by pinching or crushing it. Similar studies (Bessou and Perl, 1969) have also found that many small-diameter C fibres transmit information when the skin is damaged by intense pressure or heat; some of these receptor-fibre units have high thresholds and respond only when the skin is injured, while others have a wide response range and fire with increasingly higher frequencies as the stimulus intensity is increased from weak to noxious levels. Data such as these have led to the conclusion that specific groups of A-delta and C

fibres are 'pain fibres' – that they, and only they, are responsible for pain sensation (Perl, 1971).

There is no denying that these fibres have highly specialized
functions or that they play an important role in pain mechanisms. But there are many reasons for doubting that they are
specifically or exclusively pain fibres (Melzack and Wall,
1962, 1965). It is evident that noxious stimuli tend to excite
receptor-fibre units across the full diameter range. Because
painful stimuli are usually intense, they generally fire many
low-threshold as well as high-threshold receptor fibre units.
Collins, Nulsen and Randt (1960) stimulated peripheral nerves
in human subjects and found that pain sensation was felt only
when the electrical stimulus was sufficiently intense to fire the
small-diameter fibres. The stimulus, however, also fired the
large-diameter fibres. The number of fibres activated by a
stimulus, then, tends to increase with increasing intensity.
Furthermore, many receptor-fibre units show increased
firing rates with increasing intensity of stimulation ranging
from weak to noxious levels (Bessou and Perl, 1969). The
total number of active fibres and their rate of firing may
therefore be at least as important a determinant of pain as the
particular fibre-diameter groups that are active.

Investigators such as Bishop (1946) have proposed that a
noxious stimulus (such as a hammer blow on the thumb)
produces two pains (a fast sharp one and a slower dull one)
because of the different conduction rates in A-delta and C
fibres respectively. If the so-called 'second pain' is to be
attributed to conduction of the C fibres, and if some C fibres
also respond to light touch applied to their receptors (Douglas
and Ritchie, 1957; Bessou and Perl, 1969), then one may ask
why there are not reports of 'second touch'. Indeed the recent
evidence (Bishop, 1959) on the diameters of fibres activated by
different stimuli would have us seek two pains, two warms,
two colds and three touches. Furthermore, a large number of
receptor-fibre units exhibit 'multiple specificities': each responds to tactile, thermal *and* noxious stimuli. If these sensory
fibres are modality-specific, to which modality do they belong?

Recent data on the properties of afferent fibres (Perl, 1971) show that they are as complex as the receptors to which they are attached. There is no strict correlation between fibre size and size of receptive field, level of tonic activity, or any other property yet investigated. The evidence, however, does show that fibre diameter is correlated to some extent with the thresholds of the receptor-fibre units, and the destinations of the fibres in the spinal cord. Thus, the high-threshold receptors are associated with the small-diameter A-delta and C fibres, while the low-threshold receptors are associated with fibres that occupy the full diameter range from the large A-beta fibres to the small C fibres. Mild or moderate intensities of stimulation, therefore, activate fibres throughout the diameter range. As the intensity rises, however, an increasing number of small fibres is recruited. There is a similar relationship between fibre diameter and the destination of fibres in the central nervous system. Only the largest diameter A fibres project to the dorsal column nuclei, while the full diameter range of fibres connect with dorsal horn cells (Wall, 1961).

The original hope that each quality of cutaneous sensation is subserved by a specific fibre-diameter group has obviously not materialized. The evidence suggests that fibres within the A-delta and C groups have specialized functions that play an important role in pain; their role, however, is more subtle and complex than that implied by simply labelling them as 'pain fibres'. This role is partly revealed in recent physiological studies of spinal cord cells.

Spinal cord

Information from the body is transmitted to the brain through several spinal cord pathways: the spinothalamic tract, the dorsal column system, the dorsolateral tract (Morin's tract), and a network of short fibres designated as the propriospinal fibre system (Figures 10 and 11). All of these, directly or indirectly, may play a role in pain.

The spinothalamic tract has come to be called the 'pain tract' because anterolateral cordotomy (Figure 4, p. 52) sometimes diminishes pain. Yet the fact that this pathway carries information that may give rise to pain does not mean that it is

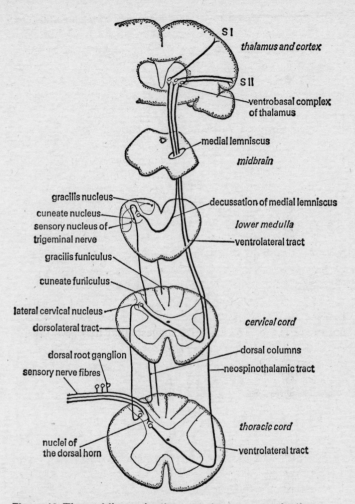

Figure 10 The rapidly conducting somatosensory projection pathways. The three main projection pathways are the dorsal column–medial lemniscal pathway, the dorsolateral tract (of Morin), and the neospinothalmic tract. The lower sections are shown on a larger scale than the upper sections. (from Milner, 1970)

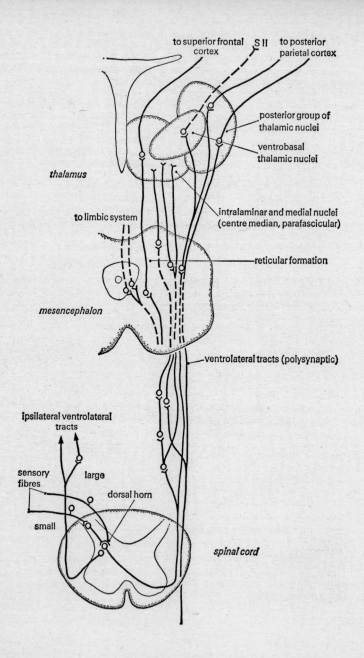

to superior frontal cortex   S II   to posterior parietal cortex

posterior group of thalamic nuclei

ventrobasal thalamic nuclei

thalamus

intralaminar and medial nuclei (centre median, parafascicular)

to limbic system

reticular formation

mesencephalon

ventrolateral tracts (polysynaptic)

ipsilateral ventrolateral tracts

sensory fibres

large

dorsal horn

small

spinal cord

the 'pain tract', with the implication that its only function is to transmit impulses involved in pain processes. Melzack and Wall (1962, p. 341) have noted that:

Surgical interruption of a pathway does much more than simply prevent particular fibres from transmitting centrally: (1) it decreases the total number of responding neurons; (2) it changes the temporal and spatial patterns of impulses still arriving at the brain; (3) it affects the descending feedback that controls transmission from peripheral fibres to dorsal horn cells in the spinal cord; (4) it alters the relationships among all the ascending sensory systems. It seems wrong, therefore, to attribute the reduction of pain after lateral tractotomy to mere blockage of modality-specific nerve impulses, without considering all the other consequences of surgical intervention.

Similarly, the fact that the dorsal column system transmits information when the skin is touched (Rose and Mountcastle, 1959) does not make it the 'touch pathway'. There is reason to believe that it may also play a role in pain. Lesions of the dorsal columns sometimes produce hyperaesthesia (Noordenbos, 1959), and may transform severe pathological itch into frank pain (Rothman, 1943). Furthermore, electrical stimulation of the dorsal columns is sometimes effective in relieving pain (Shealy, Mortimer and Hagfors, 1970; Nashold and Friedman, 1972). Noordenbos (1959), moreover, has proposed that the propriospinal system may provide yet another conduction route related to pain. Pain experience and response, then, appear to be a function of nerve impulses that ascend not one but several spinal cord pathways.

The dorsal horns (Figure 11), which receive fibres from the body and project impulses towards the brain, provide valuable clues about information processing at the spinal cord level.

Figure 11 The slowly conducting somatosensory projection pathways. The breaks in the projection lines represent multi-synaptic connections. The propriospinal fibres are not shown, but consist of short fibres which are distributed throughout the cord.
(adapted from Milner, 1970)

The dorsal horns comprise several layers or laminae, each of which is now known to have specialized functions. The inputs and outputs of each lamina are not entirely understood. But the picture which emerges, based largely on the work by Wall and his colleagues (see Hillman and Wall, 1969), reveals that the input is modulated in the dorsal horns before it is transmitted to the brain.

Lamina 1 cells are known (Christensen and Perl, 1970; Perl, 1971) to receive information from the A-delta and C fibres when the skin is crushed or burned, and a portion of them project directly to higher levels of the spinal cord. It is reasonable, therefore, to assume that they play a role in pain processes. However, the picture is far more complex than this. Just below lamina 1 are the cells that comprise the substantia gelatinosa (laminae 2 and 3). This region is of particular interest because it represents a unique system on each side of the spinal cord which appears to have a modulating effect on the input (Wall, 1964). Many afferent fibres from the skin terminate in the substantia gelatinosa, and the dendrites of many cells in lower laminae, whose axons project to the brain, lie within the substantia gelatinosa. This region, then, is situated between a major portion of the peripheral nerve fibre terminals and the spinal cord cells that project to the brain. Melzack and Wall (1965) have proposed that the substantia gelatinosa has a modulating effect on transmission from peripheral fibres to spinal cells.

Below the substantia gelatinosa, the cells in lamina 4 have small cutaneous receptive fields and project to the dorsolateral pathway ipsilaterally and probably to the lamina 5 cells (Hillman and Wall, 1969). Lamina 4 cells respond when gentle pressure is applied to the skin, and to electrical stimulation of the large A-beta myelinated fibres. However, their response rate fails to increase when the skin is pinched or crushed or when the A-delta and C fibres are activated. These cells, then, are selectively tuned to gentle pressure applied within their receptive fields.

In contrast, lamina 5 cells have a wide dynamic range and are particularly responsive when noxious stimuli are applied within their receptive fields (Hillman and Wall, 1969). Their fields have a remarkably complex organization, and they respond with characteristic firing patterns to stimulation over a wide range of intensities. Moreover, lamina 5 cells receive multiple inputs. There is reason to believe that they receive inputs from the lamina 4 cells. In addition, they receive inputs from the small myelinated and unmyelinated fibres from the skin, from deeper tissues such as blood vessels and muscles, and from the viscera (Pomeranz, Wall and Weber, 1968). Furthermore, there is convincing evidence that lamina 5 cells are under the control of fibres that descend from the brain. The lamina 5 cells, moreover, have extensive projections to the brain. The majority project through the spinothalamic tract, while some appear to project through the dorsolateral and dorsal column systems.

Hillman and Wall (1969) have shown that each cell in lamina 5 has a three-zoned receptive field (Figure 12). The cell is excited by the full range of mechanical stimuli applied to the centre of the field (zone 1), and the firing rate increases with increasing intensity of stimulation. Intense stimulation produces very prolonged repetitive discharges. Careful examination further reveals that inhibition follows gentle stimulation, and facilitation follows intense stimulation within this central area. Around this zone is a region (zone 2) where gentle tactile stimuli or electrical stimulation of large fibres produce inhibition, while intense stimuli or small-fibre stimulation produce excitation and some facilitation. These two excitatory fields are surrounded by an even larger area (zone 3) in which natural stimuli do not excite but, instead, inhibit the firing of the lamina 5 cell.

The mechanism of the inhibition produced by the large fibres and the facilitation produced by the small fibres is unknown, but Hillman and Wall (1969) suggest that it may be due to pre- and post-synaptic effects produced by the small cells of laminae 2 and 3. A similar effect has been observed by Mendell and Wall (1964). They found that a single electrical pulse delivered to small fibres produces a burst of nerve impulses followed by repetitive discharges in spinal cord cells.

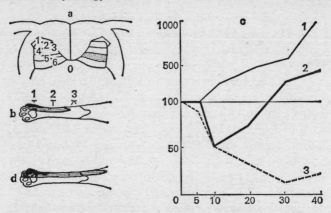

Figure 12 Receptive fields of a single cell in lamina 5 of the dorsal horns. A shows the location of lamina 5. B illustrates the receptive fields of the cell in the decerebrate cat: brushing, touching or crushing the skin in *zone 1* produced excitation of the cell; brushing in *zone 2* inhibited the cell, whereas pressure and pinch excited it; all stimuli applied to *zone 3* (outer boundaries too vague to be shown) inhibited the cell and never excited it. The effects of electrical stimulation in zones 1, 2, and 3 are shown in C: the resting firing rate of the cell is shown as 100 per cent; the stimulation voltages are indicated at the bottom. Line 1 shows the effect of stimulation in zone 1: all effective voltages produced an increase of the firing rate. Line 2 shows the effect of stimulation in zone 2; low voltages produced an inhibition and higher voltages increased the firing rate. Line 3 shows that stimulation in zone 3 produced inhibition of the cell's activity. D shows the effects of removing the brainstem influence (by cooling the upper spinal cord) on the cell: zones 1 and 2 expanded, and zone 3 disappeared.
(from Hillman and Wall, 1969, p. 284)

Successive pulses produce a 'wind-up' effect – a burst followed by a discharge of increasing duration after each stimulation. In contrast, successive pulses delivered to large fibres produce a burst of impulses followed by a 'turn-off' or period of silence after each pulse. These opposing effects of facilitation and inhibition after small and large fibre stimulation are believed (Wall, 1964) to be mediated by the substantia gelatinosa, and provide the basis of the gate-control theory described in chapter 6.

Brain mechanisms

It is traditionally assumed that pain sensation and response are subserved by a 'pain centre' in the brain. The concept of a pain centre, however, is totally inadequate to account for the complexity of pain. Indeed, the concept is pure fiction, unless virtually the whole brain is considered to be the pain centre, because the thalamus, hypothalamus, brainstem reticular formation, limbic system, parietal cortex, and frontal cortex are all implicated in pain perception. Other brain areas are obviously involved in the emotional and motor features of pain. The idea of a centre in the brain which is exclusively responsible for pain therefore becomes meaningless.

The anatomy and physiology of the somatosensory projection systems in the brain are highly complex (Figure 13). So many areas are involved in pain processes and they interact so extensively that attempts to describe them all are usually more confusing than clarifying. It is possible, instead, to begin with the major psychological dimensions of pain and to attempt to relate them to known brain structures and their functions.

The dimensions of pain

The problem of pain, since the beginning of this century, has been dominated by the concept that pain is purely a sensory experience. Yet it has a unique, distinctly unpleasant, affective quality that differentiates it from sensory experiences such as sight, hearing, or touch. It becomes overwhelming, demands immediate attention, and disrupts ongoing behaviour and thought. It motivates or drives the organism into activity aimed at stopping the pain as quickly as possible. To consider only the sensory features of pain, and ignore its motivational-affective properties, is to look at only part of the problem. Even the concept of pain as a perception, with full recognition of past experience, attention, and other cognitive influences, still neglects the crucial motivational dimension.

The motivational-affective dimension of pain is brought clearly into focus by clinical studies on frontal lobotomy,

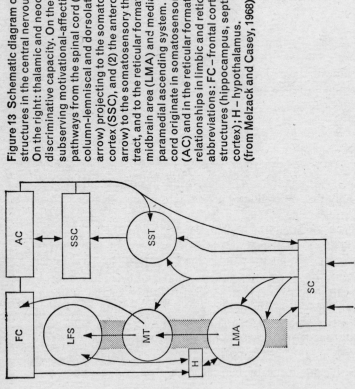

Figure 13 Schematic diagram of the major relationships among structures in the central nervous system that are related to pain On the right: thalamic and neocortical structures subserving discriminative capacity. On the left: reticular and limbic systems subserving motivational-affective functions. Ascending pathways from the spinal cord (SC) are: (1) the dorsal column-lemniscal and dorsolateral tracts (right ascending arrow) projecting to the somatosensory thalamus (SST) and cortex (SSC), and (2) the anterolateral pathway (left ascending arrow) to the somatosensory thalamus via the neospinothalamic tract, and to the reticular formation (stippled area), the limbic midbrain area (LMA) and medial thalamus (MT) via the paramedial ascending system. Descending pathways to spinal cord originate in somatosensory and associated cortical areas (AC) and in the reticular formation. Polysynaptic and reciprocal relationships in limbic and reticular systems are indicated. Other abbreviations: FC – frontal cortex; LFS – limbic forebrain structures (hippocampus, septum, amygdala, and associated cortex); H – hypothalamus.
(from Melzack and Casey, 1968)

congenital insensitivity to pain, and pain asymbolia. Patients who have undergone a frontal lobotomy (which severs the connections between the prefrontal lobes and the thalamus) rarely complain about severe clinical pain or ask for medication (Freeman and Watts, 1950). Typically, these patients report after the operation that they still have pain but it does not bother them. When they are questioned more closely, they frequently say that they still have the 'little' pain, but the 'big' pain, the suffering, the anguish are gone. It is certain that the sensory component of pain is still present because these patients may complain vociferously about pin prick and mild burn. Indeed, pain perception thresholds may be lowered (King, Clausen and Scarff, 1950). The predominant effect of lobotomy appears to be on the motivational-affective dimension of the whole pain experience. The aversive quality of the pain and the drive to seek pain relief both appear to be diminished.

People who are congenitally insensitive to pain also appear to have no sensory loss, and are able to feel pricking, warmth, cold, and pressure. They give accurate reports of increasing intensity of stimulation, but the input, even at intense, noxious levels, seems never to well up into frank pain. The evidence (Sternbach, 1968) suggests that it is not the sensory properties of the input but rather the motivational-affective properties that are absent. Similarly, patients exhibiting 'pain asymbolia' (Rubins and Friedman, 1948) after lesions of portions of the parietal lobe or the frontal cortex are able to appreciate the spatial and temporal properties of noxious stimuli (for example, they recognize pin pricks as sharp) but fail to withdraw or complain about them. The sensory input never evokes the strong aversive drive and negative affect characteristic of pain experience and response.

These considerations suggest that there are three major psychological dimensions of pain: sensory-discriminative, motivational-affective, and cognitive-evaluative. Melzack and Casey (1968) have proposed that they are subserved by physiologically specialized systems in the brain.

The sensory-discriminative dimension

Physiological and behavioural studies suggest that the sensory-discriminative dimension of pain is subserved, at least in part, by the neospinothalamic projection to the ventrobasal thalamus and somatosensory cortex (Figure 10, p. 88). Neurons in the ventrobasal thalamus, which receive a large portion of their afferent input from the neospinothalamic projection system, show discrete somatotopic organization even after dorsal column section. Studies in human patients and in animals (see Wall, 1970) have shown that surgical section of the dorsal columns, long presumed to subserve virtually all of the discriminative capacity of the skin sensory system, produces little or no loss in fine tactile discrimination and localization. Furthermore, Semmes and Mishkin (1965) found marked deficits in tactile discriminations that are attributable to injury of the cortical projection of the neospinothalamic system. These data, taken together, suggest that the neospinothalamic projection system has the capacity to process information about the spatial, temporal, and magnitude properties of the input.

The motivational-affective dimension

There is convincing evidence (Melzack and Casey, 1968) that the brainstem reticular formation and the limbic system, which receive projections from the spinoreticular and paleospinothalamic components of the anterolateral somatosensory pathway, play a particularly important role in the motivational-affective dimension of pain. These medially coursing fibres, which comprise a 'paramedial ascending system' (Melzack and Casey, 1968), tend to be short and connect diffusely with one another during their ascent from the spinal cord to the brain. They are not organized to carry discrete spatial and temporal information. Their target cells in the brain usually have wide receptive fields, sometimes covering half or more of the body surface. In addition to the convergence of somatosensory fibres, inputs from other sensory systems, such as vision and audition, also arrive at many of these cells.

Reticular formation. It is now well established that the reticular formation is involved in aversive drive and similar pain-related behaviour. Stimulation of nucleus gigantocellularis in the medulla (Casey, 1971), and the central grey and adjacent areas in the midbrain (Spiegel, Kletzkin and Szekeley, 1954; Delgado, 1955) produces strong aversive drive and behaviour typical of responses to naturally occurring painful stimuli. In contrast, lesions of the central grey or spinothalamic tract produce marked decreases in responsiveness to noxious stimuli (Melzack, Stotler and Livingston, 1958). Similarly, at the thalamic level, 'fear-like' responses associated with escape behaviour have been elicited by stimulation in the medial and adjacent intralaminar nuclei of the thalamus (Roberts, 1962). In the human, lesions in the medial thalamus (parafascicular and centromedian complex) and intralaminar nuclei have provided relief from intractable pain (Mark, Ervin and Yakovlev, 1963; White and Sweet, 1969).

There is considerable specialization of function within the reticular formation, and the effects of lesions vary from area to area. At the midbrain level, the spinothalamic tract, the central tegmental tract and the central grey pathway are all affected by analgesic agents. Responses evoked in the trigeminal component of these areas by stimulation of the tooth pulp are significantly depressed by nitrous oxide in amounts that abolish awareness of pain in human patients (Figure 14). Yet lesions in each area produce a different effect. Melzack, Stotler and Livingston (1958) found that cats with lesions of the spinothalamic tract showed a significant decrease in the capacity to perceive and respond to noxious heat and pin prick; they responded to only about 50 per cent of the stimulus presentations throughout the five testing days. Cats with central grey lesions perceived and responded to only about 30 per cent of the presentations on the first day, but improved each day until, on the fifth day, they responded to about 70 per cent of the presentations. Since the cats were tested long after the surgical wounds had healed, this change in behaviour suggests a reorganization of connections in the nervous system, as though inputs that normally travelled

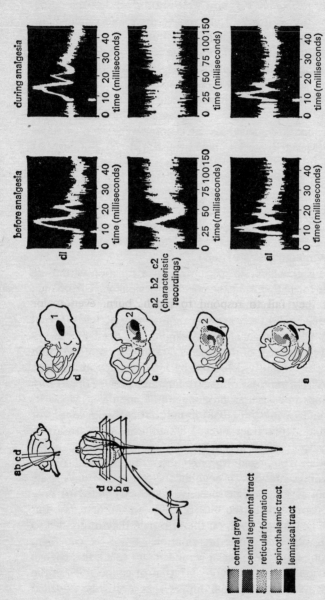

Figure 14 Five pathways in the brainstem transmit signals evoked by stimulating the nerve of a cat's tooth. The sections a, b, c and d show how the pathways progress through the midbrain and thalamus. An analgesic mixture of nitrous oxide and oxygen blocks the signals in four (2) of the five pathways. The signal is not blocked, however, in the lemniscal pathway (1). (from Haugen and Melzack, 1957, p. 183)

through one channel were now shunted through other pathways. The most surprising behavioural change, however, occurred in cats with lesions of the central tegmental tract. They became hyper-responsive to pin prick and tried to escape it with an urgency that was not observed in any of the other animals. They also showed signs of spontaneous pain: they rubbed and bit their paws, and often whined in the absence of stimulation.

These effects suggest that the central tegmental tract exerts an inhibitory effect on input. It has already been noted that electrical stimulation in or near this region produces a decrease in the size of the receptive fields of spinal cord cells. And, indeed, this region is the source of a substantial portion of descending reticulo-spinal fibres (as well as ascending fibres to thalamic and cortical projections of the somatosensory system). This inhibitory influence may help explain an exciting recent discovery. Reynolds (1969, 1970) has observed that electrical stimulation in the region of the central grey and central tegmental tract produces a marked analgesia in rats, so that they fail to respond to pinch, burn, even major abdominal surgery. These animals are not paralysed but seem instead to be completely oblivious to stimuli that are normally painful. More recently, Mayer, Wolfle, Akil, Carder and Liebeskind (1971) observed the same phenomenon, and found, moreover, that there is some somatotopic organization within the system, so that stimulation at a given site produces analgesia of only selected portions of the body, such as the lower half, or one quadrant. They found, furthermore, that the electrical stimulation of these areas appears 'pleasurable' to the animals – they actively seek it out by pressing a lever to stimulate themselves. These observations, as we shall see in chaper 6, have great theoretical significance since they suggest the presence of a system that exerts a tonic, widespread inhibitory influence on transmission through the somatosensory projection system.

Although these reticular areas are clearly involved in pain, they may also play a role in other somatosensory processes. Casey (1971) found that sixteen out of twenty cells in nucleus gigantocellularis responded to tapping or moderate pressure on the skin. The response pattern of the cells, moreover, was a function of the intensity of stimulation; the cells responded with a more intense and prolonged discharge to stimuli (pinch, pin prick) that elicited withdrawal of the tested limb. Similarly, Becker, Gluck, Nulsen and Jane (1969) found that many cells in the midbrain central grey and tegmentum responded to electrical stimulation of large, low-threshold fibres. An increase in the stimulus level in order to fire the small, high-threshold fibres produced distinctively patterned responses showing high discharge rates, prolonged afterdischarges for several seconds, and the 'wind-up' effect (increasing neural response to repeated intense stimuli).

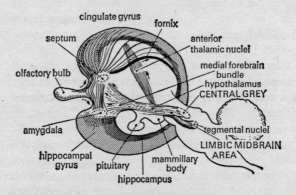

Figure 15 Schematic drawing of the limbic system, which is known to play an important role in emotional and motivational processes. The arrows indicate the direction of flow of nerve impulses through the system.
(adapted from MacLean, 1958, p. 1723)

Limbic system. The reciprocal interconnections between the reticular formation and the limbic system is of particular importance in pain processes (Melzack and Casey, 1968). The midbrain central grey, which is traditionally part of the reticular formation, is also a major gateway to the limbic system (Figure 15). It is part of the 'limbic midbrain area' (Nauta, 1958) that projects to the medial thalamus and hypothalamus which in turn project to limbic forebrain structures. Many of these areas also interact with portions of the frontal cortex that are sometimes functionally designated as part of the limbic system. Thus the phylogenetically old paramedial ascending system, which is separate from but in parallel with the newer neospinothalamic projection system, gains access to the complex circuitry of the limbic system.

It is now firmly established that the limbic system plays an important role in pain processes. Electrical stimulation of the hippocampus, amygdala, or other limbic structures may evoke escape or other attempts to stop stimulation (Delgado, Rosvold and Looney, 1956). After ablation of the amygdala and overlying cortex, cats show marked changes in affective behaviour, including decreased responsiveness to noxious stimuli (Schreiner and Kling, 1953). Surgical section of the cingulum bundle, which connects the frontal cortex to the hippocampus, also produces a loss of 'negative affect' associated with intractable pain in human subjects (Foltz and White, 1962). This evidence indicates that limbic structures, although they play a role in many other functions, provide a neural basis for the aversive drive and affect that comprise the motivational dimension of pain.

Intimately related to the brain areas involved in aversive drive, and sometimes overlapping with them, are hypothalamic and limbic structures that are involved in approach responses and other behaviour aimed at maintaining and prolonging stimulation (Olds and Olds, 1963). Electrical stimulation of these structures often yields behaviour in which the animal presses one bar to receive stimulation and another to stop it. These effects, which may be due to overlap of 'aversive' and 'reward' structures, are sometimes a function simply of intensity of stimulation, so that low-level stimulation elicits approach and intense stimulation evokes avoidance. Complex interactions among these areas (Olds and Olds, 1962) may explain why aversive drive to noxious stimuli can be blocked by stimulation of reward areas in the lateral hypothalamus (Cox and Valenstein, 1965).

These data show clearly that the neural areas comprising

the paramedial, reticular, and limbic systems are involved in the motivational and affective features of pain. The manner in which these areas are brought into play will be discussed in chapter 6.

The cognitive-evaluative dimension

We have already seen (chapter 2) that cognitive activities such as cultural values, anxiety, attention and suggestion all have a profound effect on pain experience. These activities, which are subserved in part at least by cortical processes, may affect the sensory-discriminative dimension or the motivational-affective dimension. Thus, excitement in games or war appears to block both of these dimensions of pain, while suggestion and placebos may modulate the motivational-affective dimension and leave the sensory-discriminative dimension relatively undisturbed.

Cognitive functions, then, are able to act selectively on sensory processing or motivational mechanisms. In addition, there is evidence that the sensory input is localized, identified in terms of its physical properties, evaluated in terms of past experience, and modified *before* it activates the discriminative or motivational systems. Men wounded in battle may feel little or no pain from the wound but may complain bitterly about an inept vein puncture (Beecher, 1959). Dogs that repeatedly receive food immediately after the skin is shocked, burned, or cut soon respond to these stimuli as signals for food and salivate, without showing any signs of pain, yet howl as normal dogs would when the stimuli are applied to other sites on the body (Pavlov, 1927, 1928).

The neural system that performs these complex functions of identification, evaluation, and selective input modulation must conduct rapidly to the cortex so that somatosensory information has the opportunity to undergo further analysis, interact with other sensory inputs, and activate memory stores and pre-set response strategies. It must then be able to act selectively on the sensory and motivational systems in order to influence their response to the information being transmitted over more slowly conducting pathways. Melzack and

Wall (1965) have proposed that the dorsal-column and dorso-lateral projection pathways (Figure 10, p. 88) act as the 'feed-forward' limb of this loop. The dorsal column pathway, in particular, has grown apace with the cerebral cortex (Bishop, 1959), carries precise information about the nature and location of the stimulus, adapts quickly to give precedence to phasic stimulus changes rather than prolonged tonic activity, and conducts rapidly to the cortex so that its impulses may begin activation of central control processes.

Influences that descend from the cortex are known to act, via pyramidal and other central-control fibres, on portions of the sensory-discriminative system such as the ventrobasal thalamus (Shimazu, Yanagisawa and Garoutte, 1965). More-over, the powerful descending inhibitory influences exerted on dorsal-horn cells in the spinal cord (Hagbarth and Kerr, 1954; Hillman and Wall, 1969) can modulate the input before it is transmitted to the discriminative and motivational systems (Figure 16). These rapidly conducting ascending and descend-ing systems can thus account for the fact that psychological processes play a powerful role in determining the quality and intensity of pain.

The frontal cortex may play a particularly significant role in mediating between cognitive activities and the motivational-affective features of pain (Melzack and Casey, 1968). It receives information via intracortical fibre systems from virtually all sensory and associational cortical areas and projects strongly to reticular and limbic structures. The effects of lobotomy, which are characterized by lowered affect and decreased drive for narcotics and other methods of pain relief, could be due to a disruption of the regulating effects of central control processes on activity in the reticular and limbic sys-tems.

Spatial and temporal patterning

We have already noted that somatic sensory inputs act on an organism that has mechanisms, both innate and acquired, for input selection and abstraction. Perception and response involve the classification of the multitude of patterns of nerve

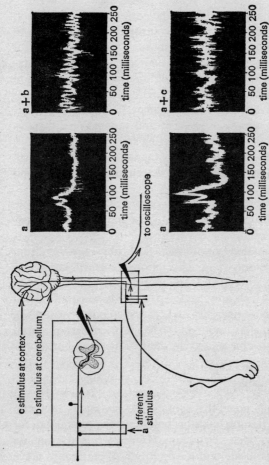

Figure 16 Effects of stimulation of brain structures on transmission in the spinal cord. An afferent nerve entering the spinal cord is electrically stimulated directly (a). The signal passes through the dorsal horn cells and is recorded on the other side of the cord (as it ascends to the brain), producing tracing a. If the cerebellum (b) or the cortex (c) is stimulated simultaneously, the afferent signal is almost completely suppressed, as shown in tracings a + b and a + c.
(from Hagbarth and Kerr, 1954, p. 295)

impulses arriving from the body, and are a function of the capacity of the brain to select and to abstract from the total information it receives from the somatosensory system (Head, 1920). 'Modality' classes are just such abstractions from the information conveyed by the entire somatosensory system.

To say that information is coded in the form of spatial and temporal patterns of nerve impulses is not enough. We need to know the particular information-codes that are sent to the brain when the skin is stimulated, and the way central cells select from this information to provide the many different qualities of our sensory experience. Mechanisms for the detection of the *spatial* and *temporal* properties of the input are believed to exist at all levels of the central nervous system, so that information is selected at several levels successively and even in parallel systems at the same time (Melzack and Wall, 1962).

Spatial patterning

Information is conveyed by the spatial patterns of nerve impulses – that is, the activity patterns in many neurons. There are two aspects of spatial patterns that are especially relevant for understanding pain: *convergence* of many neurons on to one cell, and *divergence* of neurons to different structures.

Convergence may play a particularly important role in pain. Many receptor-fibre units from the skin may converge onto a single cell in the spinal cord. And many of these, in turn, may converge onto a single cell in the brain. These converging inputs provide the conditions for spatial summation. We have already noted that the number of fibres activated by a stimulus tends to increase with increasing stimulus intensity. Intense pressure, thermal or chemical stimulation will fire not only high-threshold fibres but many fibres throughout the full lower range. Experiments that correlate loss of pain sensation with selective block of C fibres are usually interpreted to mean that C fibres are pain fibres (Bishop, 1959). However, it seems more likely (Melzack and Wall, 1965) that pain results when the total afferent barrage in all fibres exceeds a critical level and that the only way to exceed the level is by activation of small-diameter fibres. Pain, then, as many writers (Livingston, 1943; Hebb, 1949; Noordenbos, 1959) have suggested, may be due to central cells responding in a particular way to the number of active fibres and the frequency of impulses per fibre.

In addition to convergence, there is also a *divergence* of fibres to different brain areas which have specialized functions. We have already noted that some areas of the central nervous system appear to have specialized sensory-discriminative functions, while other areas seem to be particularly involved in motivational-affective processes. The divergence of afferent fibres going to the dorsal horn cells and the dorsal column nuclei marks only the first stage of the process of selection of information. The fibre systems that arise at the dorsal horns diverge again at higher levels of the central nervous system and give rise to ascending pathways which have different functions. Indeed, the pathways may recombine and diverge a number of times, to allow a more precise analysis of the input. This analysis, moreover, can be influenced by activity in descending fibres which impinge on central cells at several levels. All of these activities must go on at the same time, yet underlie the observed unity of experience and response.

Temporal patterning

Information is conveyed by the temporal patterning of nerve impulses in individual neurons. Single cells in the spinal cord, for example, show distinctive firing patterns when they are activated by stimuli applied to their receptive fields at the skin. Wall (1960) and Wall and Cronly-Dillon (1960) have recorded patterns to hair movement, touch, temperature and skin-damage that are distinctively different and appear to provide a temporal code (Figure 17). The code, of course, must be 'read' by cells in the later stages of information selection, and Melzack and Wall (1962) have speculated on the type of temporal pattern which might be recognized by central cells.

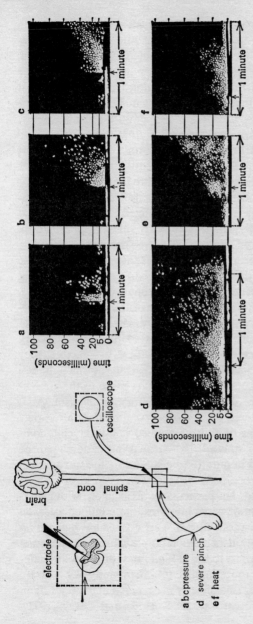

Figure 17 Neuron firing patterns, recorded from single cells in the spinal cord of a cat, show the initial response in the central nervous system to various stimuli applied to the cat's leg. Pattern a was caused by hanging a two-gram weight on a single hair; b shows the effect of a twenty-gram weight; c is the effect of a mild pinch. All three stimuli start at the arrows and continue for the duration of the recording. In d the skin was severely pinched for one minute. In e and f a heat lamp was directed at the skin for fifteen seconds after the arrows, raising the skin temperature four and twelve degrees centigrade respectively. Each dot in the recordings represents a single nerve impulse; height above base line represents time interval between recorded impulse and preceding one. (from Wall, 1960, and Wall and Cronly-Dillon, 1960, p. 365)

The first temporal pattern is the rapid, sudden increase and decrease of firing-frequency characteristic of fibres that respond to gentle pressure. Central cells with high threshold, rapid adaptation, or post-excitatory inhibition would be tuned to respond only to this temporal pattern. The next pattern is the prolonged, relatively steady, low rate of bombardment characteristic of the temperature sensitive fibres. Central cells with a low threshold, slow response, and no adaptation would be tuned to this time pattern. The last temporal pattern is the rapid rise of frequency, followed by slow fall, characteristic of receptor-fibre units that respond to heavy pressure or tissue damage. Central cells of high threshold which respond only after prolonged bombardment and do not adapt rapidly would be tuned to this temporal pattern.

Wall and Cronly-Dillon (1960) have speculated further on the physiological properties of cells that are attuned to intense inputs. They note that as the afferent barrage rises, the output response of these cells also rises. Two physiological properties, however, influence the firing patterns of these cells: the time-constant of the excitation and after-discharge from afferent fibres; and the longer time-constant of the post-excitatory inhibition of the central cell. Given these two factors, they predict that there is a level of afferent barrage at which the cell will oscillate between high frequency response to the afferent barrage and silence due to the inhibitory mechanism. And, indeed, they observed bursts of nerve impulses after they crushed small areas of skin or applied itch-producing compounds. They also showed that these firing patterns can be altered and limited in duration by subjecting the surrounding skin to a vibratory stimulus (Figure 18).

Pain appears to be particularly associated with after-discharges – volleys of nerve impulses that continue after the cessation of stimulation. Spinal cord cells show high firing rates for the duration of stimulation (such as pinching the skin) and may continue to fire at high rates for several minutes after removal of the stimulus (Hillman and Wall, 1969). Comparable phenomena have been observed in the brainstem. Becker, Gluck, Nulsen and Jane (1969) and Casey (1971),

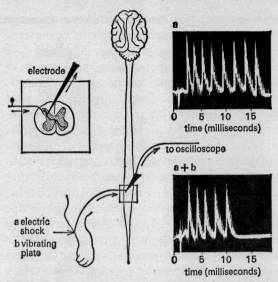

Figure 18 Modification of nerve signals by vibration. A single electric shock at the skin produces a train of nerve impulses (a) in a neuron in the spinal cord. The impulse train is shortened when the skin around the shocked region is vibrated by a metal plate (b). (from Wall and Cronly-Dillon, 1960, p. 365)

recording from cells in the brainstem reticular formation, observed that noxious stimulation of the skin produces high frequency firing followed by after-discharge bursts for hundreds of milliseconds after cessation of stimulation.

There is a further pattern that may be related to pain. It was noted in chapter 3 that several pathological pain states are associated with partial destruction of peripheral sensory nerves. Many of the neuralgias occur after partial nerve injury, and causalgia occurs more frequently after partial than after total peripheral nerve lesions (Sunderland, 1968). Even phantom limb pain is associated with partial nerve damage since only a portion of the fibres in a nerve bundle degenerate after total section, while the remainder tend to regenerate into the stump tissues. It is of interest, then, to consider the effects of partial deafferentation of nerves on the activity of cells in the spinal cord.

Ward (1969) observed that cutting one or more sensory spinal roots in the cat produces abnormal rhythmic bursts of firing in dorsal horn cells that persist for more than thirty days after the root section. Furthermore, single shock pulses to adjacent intact roots produce prolonged firing that persists for hundreds of milliseconds. These abnormal patterns are seen only after partial deafferentation. Some input, in other words, is necessary to trigger the rhythmic bursts. Similar abnormal activity in trigeminal cells in the cat is also seen after extraction of all the teeth on one side. It has also been observed in human spinal cord cells from which records have been made after dorsal root section. It is of particular interest that this abnormal activity has been recorded from dorsal horn cells that include lamina 5. Lamina 5 cells, as we have already seen, are believed to be involved in sensory transmission processes related to pain experience and response. Whether or not these bursts are in fact related to pain is a matter of conjecture. However, they provide a tantalizing bit of information that may be related to pain states that are characterized by degeneration of sensory nerve fibres or dorsal root pathology.

Patterning and pain

The concept that pain is associated with particular patterns of sensory input has led investigators to see whether changes in the patterns bring about changes in pain experience.

Pain and vibration. It was noted earlier that vibration of the skin modifies the firing patterns evoked in spinal cord cells by noxious stimuli applied to adjacent skin areas. There is now evidence (Wall and Cronly-Dillon, 1960; Melzack, Wall and Weisz, 1963; Melzack and Schecter, 1965) that vibration is also able to modify the amount of pain produced by noxious stimulation of the skin in normal subjects. In general, vibration decreases the perceived intensity of mild or moderate levels of pain, but makes high levels of pain even worse. The opposite, however, seems to occur in pathological pain syndromes. Causalgic and neuralgic pains are enhanced by mild stimulation, so that even gentle touches may evoke excruciating pain. More powerful stimulation, such as pinch, heavy pressure or vigorous massage, usually fails to evoke pain and may instead produce pain relief. Even pain-producing stimuli may relieve pathological pain. The effect of one input pattern on another, then, appears to depend on the particular patterns involved.

The timing of the interaction is of special interest. Melzack, Wall and Weisz (1963) found that a slap on the skin decreases the level of perceived pain when it either precedes or follows an electric shock by as long as fifty milliseconds. Halliday and Mingay (1961), furthermore, have found that the threshold for shock on one arm is raised by a shock delivered as long as a hundred milliseconds later to the other arm. This phenomenon, which is known as 'metacontrast', is powerful evidence that pain results after prolonged monitoring of the input pattern by central cells. It is assumed that this monitoring may take even longer in pathological pain syndromes (chapter 3), since pain often occurs after delays as long as thirty seconds or more after stimulation. Such long delays simply cannot be explained in terms of slowly conducting fibres.

These data also indicate that the pain-signalling pattern can be influenced by stimulation at distant body sites. Shock applied to one arm influences pain felt in the other (Halliday and Mingay, 1961). Vibration of one wrist influences the level of itch perceived at the opposite wrist (Melzack and Schecter, 1965). Care was taken in both of these studies to rule out distraction of attention as the cause of the effect; rather the changes in pain or itch appear to reflect temporal and spatial interactions of somatic inputs in the central nervous system.

Pain and stimulus size. It is evident from the study of pathological pain syndromes that pain intensity increases when the area that is stimulated is increased (Noordenbos, 1959). Curiously, such summation effects are difficult to demonstrate in normal skin, perhaps because a larger stimulus area activates the inhibitory fields that surround the peak excitatory areas, so that additional excitation is cancelled out by a concomitant increase in inhibition. Paradoxically, pain sometimes occurs more often after stimulation of small rather than large areas of skin. Head (1920) observed that immersion of the tip of the penis into water at 45°C produces pain, but that the pain disappears completely when the penis is immersed more deeply. Similarly, stimulation of the skin with a small warm stimulator (with a tip diameter of one to two millimetres) often produces reports of stinging pain, even though stimulation of a larger area of skin with a probe of the same temperature produces only reports of warmth sensation (Melzack, Rose and McGinty, 1962). These observations indicate that the spatial patterns of excitation produced by stimuli play an important role in determining the quality of cutaneous experience.

This effect is of particular interest. Touching the skin, especially the lip, with a hair or thread frequently sets off a tingling or 'afterglow' sensation that may persist for several minutes (Melzack and Eisenberg, 1968). The afterglow sometimes spreads beyond the site of stimulation, and occasionally the mirror-area on the other side of the lip may begin to tingle. There is characteristically a delay in the onset of the afterglow. These effects are even more pronounced when a small-diameter warm probe is used to stimulate the skin. The afterglow appears after a delay, wells up into sharp stinging pain, and persists long after stimulation. These effects are not found uniformly across the skin, but only in particular regions, which may vary in location from one testing period to the next. The delay, spread, after-sensations, and unpleasant sensory qualities are reminiscent of the properties of the neuralgias.

A further observation is of special relevance to pain. Many subjects report that, instead of an afterglow, or once the afterglow has ceased, they are 'aware' of the area in a unique way: some report simply the feeling of a zone of heightened awareness, while others report pulsations from the area which, they believe, reflect their breathing rate or pulsing blood vessels in the skin. The subjects report that the 'zone of awareness' subsides slowly, although some can later evoke it by intense concentration on the area. This intriguing report of a 'zone of awareness' suggests that cognitive processes, such as attention, may selectively facilitate the transmission of all inputs from a particular skin area. The pulsatile property of the experience reported by some subjects indicates that it may reflect, in part, input from blood vessels or surrounding tissue.

Two possible mechanisms may underlie the stinging pain that is sometimes evoked by stimulation with a small-tipped warm probe. First, the small stimulating surface may occasionally activate the excitatory centres of receptive fields with minimal activation of their inhibitory surrounds. The decreased inhibition would permit maximal transmission of impulses, without the inhibitory 'turn-off'. The ensuing prolonged summation could bring about stinging pain of long duration. Second, it is possible that when insufficient or ambiguous cutaneous information is applied to the skin, the tonic inhibitory control exerted by the brain is decreased so that the input undergoes maximal summation and is perceived as pain. Whatever the mechanism, it is apparent that brief stimulation of small areas of the skin may produce changes in central neural activity that continue as long as several minutes after cessation of stimulation.

Prolonged activity in the nervous system

The persistence of pathological pain (described in chapter 3) in the absence of any obvious somatic stimulation suggests that prolonged changes occur in the central nervous system. Nathan (1962) has reviewed several cases which indicate that somatosensory input may produce long-lasting effects similar to memories produced by visual and auditory stimulation. In one case, stimulation of the stump of an amputee who, five years before amputation, had sustained a severe laceration of the leg by an ice skate, later produced vivid imagery of the pain of the skating accident: 'It was not that he remembered having had this injury; he felt all the sensations again that he had felt at the time.'

The concept of a memory-like mechanism in pain is also supported by convincing experimental evidence obtained in both man and animals. After teeth on both sides of the mouth were drilled and filled without local anaesthetic, pin-pricks of the nasal mucosa, as long as seventy days later, produced pain in the treated teeth on the stimulated side. The effect was permanently abolished on one side by a single novocaine block of the trigeminal nerve, but persisted in the opposite, non-blocked side (Hutchins and Reynolds, 1947; Reynolds and Hutchins, 1948). These referred pains necessitate the assumption of a longer-term central neural change. The data suggest that the treatment of the teeth evoked inputs that produced changes in firing patterns in the central nervous system. These changes, once initiated, were somehow capable of summating the continuous, low-level input from the treated teeth with inputs from more distant sources. The single block of a peripheral nerve, which could not have affected the teeth, permitted resumption of normal neural activity and the end of pain. That the *input* as such, rather than conscious awareness, was essential in initiating the abnormal central activity is evident in the observation that a subject who had four teeth extracted under nitrous oxide anaesthesia felt pain referred to the jaw when the nasal mucosa was pricked thirty-three days after treatment.

Similar observations were made by Cohen (1944) who studied patients that had anginal-effort syndrome, with pain referred only to the left side. He injected a small amount of hypertonic saline under the skin of the right side of the back which gave rise to a diffuse, deep-seated pain that soon disappeared. Two hours later, long after the pain had passed, exertion and anginal pain again caused its appearance.

Comparable conclusions can be drawn from experiments with animal subjects. Injection of turpentine under the skin of a cat's paw produces a temporary inflammation and a tendency to flex the paw. After the inflammatory process has healed completely and the animal walks normally, an abnormal flexion–extension pattern is seen in the limbs when the animal is decerebrated (Frankstein, 1947). Similarly, postural asymmetries produced by cerebellar lesions persist after transection of the spinal cord only if they are maintained for at least forty-five minutes before the cord section (Chamberlain, Halick and Gerard, 1963). Impulses descending from the cerebellum for forty-five minutes or longer thus appear to bring about a permanent change in spinal neuron networks.

Prolonged spinal activity. There is now more direct evidence of prolonged changes in spinal cord activity. Spencer and April (1970) have shown that a seven-minute period of intense, high-frequency 'tetanic' electrical stimulation of sensory fibres produces a change – post-tetanic potentiation – in a simple spinal cord reflex for more than ten minutes after stimulation is stopped (Figure 19). Twenty minutes' stimulation produces potentiation for more than two hours. Potentiation lasting an hour or more occurs even with stimulus frequencies as low as one hundred pulses per second for twenty to thirty minutes. Since injury may produce intense, high-frequency sensory input that lasts for several hours or days, a change in central neural activity of much longer durations can be extrapolated.

Prolonged brain activity. There is also evidence of prolonged activity in brain tissue. Melzack, Konrad and Dubrovsky (1968, 1969) have observed that brief (ten to twenty seconds) stimulation of the skin or sciatic nerve in moderately anaesthetized cats produces marked, prolonged changes in the tonic, spontaneous activity at several synaptic levels of the skin sensory system, the brainstem reticular formation, and other discrete brain areas (Figure 19). The changes usually last for five to ten minutes, but may continue for as long as thirty or forty minutes before activity returns to normal. Another brief stimulation again triggers a prolonged change. These prolonged changes occur only when the animal is moderately anaesthetized but not when it is too light or too deep. The data suggest that the reticular formation plays an important role in the mechanisms underlying the prolonged changes. The most reasonable explanation of these long-lasting changes is that an inhibitory influence, which normally shuts off stimulus-evoked activity in a sensory system as soon as the stimulus is removed, is selectively abolished by anaesthetic agents. After removal of inhibition, a stimulus which usually has only momentary effects now produces persistent activity.

Possible neural mechanisms. The mechanisms that underlie the prolonged changes that have just been described are unknown. There is anatomical and physiological evidence, however, that allows us to speculate on two possible mechanisms:

1. The closed, self-exciting circuit (Figure 20) that has been

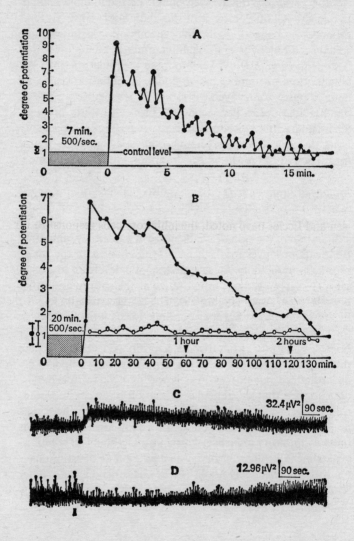

described anatomically by Lorente de Nó (1938) has served as the basis of Livingston's (1943) and Hebb's (1949) concepts of permanent memories of prior sensory experience. Although self-exciting circuits of enormous complexity have been described, it is doubtful that nerve impulses can 'reverberate' through the loops for the known time durations of memories – a half century or more. For this reason, Hebb (1949, 1972) has proposed that permanent changes occur at the synapses of these circuits that 'consolidate' the memory. It is possible, then, that prolonged, intense, or otherwise abnormal sensory inputs may bring about permanent memories in the nervous system. Inhibitory mechanisms may normally prevent the expression of these memories. Release from inhibition may activate them.

2. Recent physiological studies suggest an even simpler kind of neural circuit that can produce long-lasting activity. Andersen and Eccles (1962) and Burke and Sefton (1966) have found evidence for a two-neuron closed circuit, so that a single volley of impulses is capable of producing rhythmic, sustained activity for prolonged periods of time (Figure 20). As Andersen and Eccles have noted, the inhibitory cells responsible for

Figure 19 A, B: Spencer and April's (1970) observations on the effects of prolonged tetanizing electric shock of the tibial nerve on the post-tetanic potentiation of a monosynaptic reflex recorded from a lumbar (L7) ventral root in the acute spinal cat. The shock was applied for seven minutes (A) and twenty minutes (B). Each plotted point in A represents the amplitude (as a multiple of the control level) of the monosynaptic reflex response to test stimuli at twenty-second intervals. Each plotted point in B is an average of ten to twenty responses to test stimuli delivered at ten-second intervals. The open circle plot in B shows the response of the contralateral control. C, D: Melzack, Konrad and Dubrovsky's (1969, p. 416) observations of prolonged changes in central neural activity produced by brief stimulation. The duration of stimulation is indicated by the black bar under each record. C shows a prolonged change in the multi-unit (mean-square) activity in the medial lemniscus produced by electrical stimulation of the midbrain reticular formation. D shows a prolonged change in the ventrobasal nucleus of the thalamus produced by pinching a hindpaw.

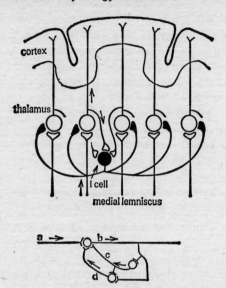

Figure 20 Models to explain long-term changes in central neural activity. *Top* shows model proposed by Andersen and Eccles (1962) to account for repetitive, rhythmic bursts of activity in the ventrobasal thalamus. Nerve impulses arriving along lemniscal fibres activate the thalamic neurons that (a) project to cortex and (b) send axon collaterals to an inhibitory neuron (I cell) that projects back to the thalamic cell bodies. The thalamic cells are inhibited briefly, then (after disinhibition and rebound excitation) fire spontaneously, reactivating the recurrent inhibitory loop. This repetitive activity within the closed loop could continue for prolonged periods of time in the absence of any further input. *Bottom* shows model of a closed, self-exciting, multi-neuron chain derived from anatomical observations by Lorente de Nó (1938, p. 207). Neuron A fires neuron B which, in turn, fires neurons C and D; C and D fire B again. This 'reverberatory' activity could bring about permanent synaptic changes that are assumed to be the basis of long-lasting 'memories'.

the self-sustaining activity could have widespread connections, producing rhythmic discharges in adjacent neuron pools and, with time, increasing the number of rhythmically firing neurons. Inputs that arrive out of phase with the rhythmic activity can block it. This type of circuit could underlie prolonged, yet reversible changes in neural activity such as those observed in the spinal cord or brain. It could also play a role, as we shall see in chapter 6, in prolonged pathological pain which is permanently abolished by injections of substances that block or intensify the sensory input.

Response mechanisms

Until now we have dealt primarily with the input properties of pain – that is, with information processing in the somatosensory projection systems. The response output is traditionally given little attention, as though once the pain alarm system is activated, impulses are transmitted directly to muscles to bring about 'pain responses'. The neural mechanisms underlying the motor aspects of pain, however, are at least as complex as those on the input side. Melzack and Wall (1965, p. 979) have noted that:

Sudden, unexpected damage to the skin is followed by (1) a startle response; (2) a flexion reflex; (3) postural readjustment; (4) vocalization; (5) orientation of the head and eyes to examine the damaged area; (6) autonomic responses; (7) evocation of past experience in similar situations and prediction of the consequences of the stimulation; (8) many other patterns of behaviour aimed at diminishing the sensory and affective components of the whole experience, such as rubbing the damaged area, avoidance behaviour, and so forth.

This total complex sequence is hidden in the simple phrase 'pain response'. Yet it is obvious that the sequence is determined by sensory, motivational, and cognitive processes acting on motor mechanisms which are known to be intricately organized throughout the brain and spinal cord (Yahr and Purpura, 1967). Melzack and Bridges (1971) have recently proposed a model to account for the selection of a particular response programme (Figure 21). Let us consider what happens, for example, when a man is bitten by a snake. The sensory, motivational, and cognitive activities at the moment would contribute to the selection of a small number of possible response programmes, such as searching for a piece of wood to kill the snake and thereby determining whether or not it is poisonous, immediate application of pressure to squeeze out any possible poison, running towards home or to the nearest hospital, and so forth. The neural activities that represent these response programmes would have the role of selectively activating (or facilitating) neural circuits for subsequent input and output processing related to the programmes. The par-

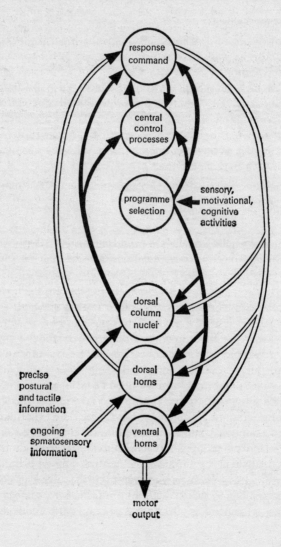

ticular programme that is selected, however, must be determined by precise postural and tactile information prior to response, and by cognitive processes that evaluate the outcome of the different response strategies. The dorsal column projection system (Figure 21) could play a role in both of these activities.

Melzack and Bridges (1971) propose that the final response command (that is, activation of a specific response programme) occurs as a result of interactions of the postural and tactile input, the central neural processes underlying evaluation of response strategies in terms of prior experience, and the active neural circuits representing several possible programme sequences (Figure 21). After the programme is activated, impulses descending the largest pyramidal fibres could selectively facilitate all the motorneuron pools necessary to carry out the entire sequence (Lashley, 1951). These impulses would be below firing threshold for the pools (Milner, 1970). Impulses descending the slower pyramidal and extrapyramidal fibres

Figure 21 Schematic model of the neural mechanisms that underlie purposive motor behaviour. Sensory, motivational, and cognitive activities at a given moment selectively activate neural circuits (Programme Selection) that represent a small number of alternative behaviour programmes to attain a goal. The output of the programmes is projected to areas at several levels of the nervous system. Neuron pools in the dorsal column nuclei are selectively facilitated, and project precise postural and tactile information to (a) areas (Central Control Processes) that evaluate the selected programmes in terms of prior experience, and (b) areas that activate the final response command. Motorneuron pools in the ventral horns, which have been facilitated by the earlier programme selection to produce several possible behaviour patterns, are selectively activated by the large and small efferent fibres, and produce the final pattern to muscles. The efferent activity to the dorsal column nuclei and the dorsal horn cells may continue to modulate the input during the ensuing motor behaviour. Open lines represent the traditional S–R circuit. Solid lines represent the circuits involved in programme selection prior to the central response command that produces overt behaviour.
(from Melzack and Bridges, 1971, p. 53)

would then provide the additional excitation to fire the pools in serial order and produce the sequential muscle patterns for overt behaviour.

Implications of the physiological evidence

The physiological evidence shows that the receptors, fibres, and central nervous system pathways involved in pain are specialized to generate and transmit patterned information rather than modality-specific impulses. Injurious stimuli activate multiple fibre systems which converge and diverge a number of times so that the patterning can undergo change at every synaptic level. Nerve impulses in large and small fibres that converge onto the cells of the dorsal horns are subjected to modulation by the activity of the substantia gelatinosa. Similarly, the convergence of fibres onto cells in the reticular formation permits a high degree of summation and interaction of inputs from spatially distant body areas. Divergence also occurs: fibres fan out from the dorsal horns and the reticular formation, and project to different parts of the nervous system that have specialized functions. One of these functions is the ability to select and abstract particular kinds of information from the temporal patterns that are conveyed by the incoming fibres. Central cells, it is now also apparent, monitor the input for long periods of time. The after-discharges, and other prolonged neural activity produced by intense stimuli, may persist long after cessation of stimulation, and may play a particularly important role in pain processes.

This convergence and divergence, summation and pattern discrimination all go on in a dynamically changing nervous system. Stimuli impinge on sensory fields at the skin that show continuous shifts in sensitivity. Furthermore, fibres that descend from the brain continually modulate the input, facilitating the flow of some input patterns and inhibiting others. The widespread influences of the substantia gelatinosa and the reticular formation, which receive inputs from virtually all of the body, can modify information transmission at almost every synaptic level of the somatosensory projection systems.

These ascending and descending interactions present a picture of dynamic, modifiable processes in which inputs impinge on a continually active nervous system that is already the repository of the individual's past history, expectations and value systems. This concept has important implications: it means that the input patterns evoked by injury can be modulated by other sensory inputs or by descending influences, which may thereby determine the quality and intensity of the eventual experience.

The somaesthetic system is a unitary, integrated system comprised of specialized component parts. Several parallel systems analyse the input simultaneously to bring about the richness and complexity of pain experience and response. Some areas are specialized to select sensory-discriminative information while others play specialized roles in the motivational-affective dimension of pain. These parallel information-processing systems interact with each other, and must also interact with cortical activities which underlie past experience, attention, and other cognitive determinants of pain. These interacting processes produce the myriad patterns of activity that subserve the varieties of pain experience.

The physiological data, taken together, demonstrate that we cannot assume a fixed, direct relationship between physical and psychological dimensions, in which all sensory differentiation occurs at the receptor level. The recent evidence on somaesthesis supports the concept proposed by Head (1920, p. 745) more than fifty years ago:

Between the impact of a physical stimulus on the peripheral endorgans of the nervous system, and the simplest changes it evokes in consciousness, lie the various phases of physiological activity. The diverse effects produced on the living organism ... must be sorted and regrouped; some are facilitated, others are repressed .... This process is repeated throughout the central nervous system until the final products of integration ... excite those conditions which underlie the more discriminative or more affective aspects of sensation .... Qualities such as pain, heat and cold are abstracted from the psychical response and spoken of as primary 'sensation';

but they have no exact physiological equivalent in the vital reactions of the peripheral mechanism. A 'primary sensation' is an abstraction. Afferent physiological processes are most complex at their origin; they become continuously more specific and simpler as they are subjected to the modifying influence of the central nervous system.

5 The Evolution of Pain Theories

So far, we have been concerned primarily with experimental and clinical observations related to pain. Some of the data, as we have seen, are still surrounded by controversy, and the scientist often has to sift the genuine facts from those 'clues' which may only lead into blind alleys. But facts alone, scientists have discovered over the centuries, usually fall short of providing a complete understanding of difficult problems. Books have been written which bring together all the known facts about pain, yet the puzzle persists. There are still too many fundamental questions for which we have no answers. Nevertheless we grope towards understanding and, for that reason, invent theories that bring us closer to it.

Although the notion of a scientific theory sounds formidable, a theory is primarily an attempted solution to a puzzle or problem – not unlike a guess made by a detective presented with an array of clues in a mystery. Several clues may lead to a theory or guess on the nature of the solution. The theory, in turn, may lead to a search for further clues that were previously not evident.

A theory alone, however, may not be enough to convince (or convict). New facts are tested against the theory to see whether or not they fit. If they support the theory, all the clues may fit together to make a coherent picture. Sometimes they demand alterations of the theory. At other times they are so much at odds with the theory that it must be rejected. In this chapter we will examine and evaluate the theories of pain that have evolved during the past century.

Specificity theory

The traditional theory of pain is known as 'specificity theory'. It is described in virtually every textbook on neurophysiology, neurology and neurosurgery, and is so deeply entrenched in medical school teaching (until recently, at least) that it is often taught as fact rather than theory. It is presented as though we already have the major answers to pain problems, and all that remains are a few minor questions that deal with therapy. It also proved to be a very powerful theory during the first half of this century, giving rise to excellent research and to some effective forms of treatment. It has several basic flaws, however, and new, more powerful theories have recently been proposed.

Specificity theory proposes that a specific pain system carries messages from pain receptors in the skin to a pain centre in the brain. To understand the theory, we must first consider its origins. The best classical description of the theory was provided by Descartes in 1644, who conceived of the pain system as a straight-through channel from the skin to the brain. He suggested that the system is like the bell-ringing mechanism in a church: a man pulls the rope at the bottom of the tower, and the bell rings in the belfry. So too, he proposed (Figure 22), a flame sets particles in the foot into activity and the motion is transmitted up the leg and back and into the head where, presumably, something like an alarm system is set off. The person then feels pain and responds to it. Because of this kind of analogy, specificity theory is also known as the alarm-bell theory or push-button theory. Despite its apparent simplicity, the theory involves several major assumptions, which we will examine shortly. First, however, we will see how Descartes' theory has evolved in the last three centuries.

The theory underwent little change until the nineteenth century, when physiology emerged as an experimental science. The problem faced by sensory physiologists in the nineteenth century was this: how can we account for the different qualities of sensation? Our visual and auditory sensations are qualitatively different from each other, just as our skin

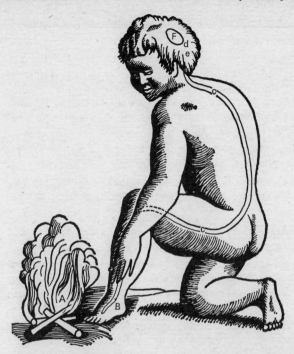

Figure 22 Descartes' (1644) concept of the pain pathway. He writes: 'If for example fire (A) comes near the foot (B), the minute particles of this fire, which as you know move with great velocity, have the power to set in motion the spot of the skin of the foot which they touch, and by this means pulling upon the delicate thread (cc) which is attached to the spot of the skin, they open up at the same instant the pore (d e) against which the delicate thread ends, just as by pulling at one end of a rope one makes to strike at the same instant a bell which hangs at the other end.'

sensations are obviously different from those of taste or smell. What is the basis of these different sensory qualities? As a result of studies by early anatomists and physiologists, it became apparent that the brain is aware of the outside world only by means of messages conveyed to it by the sensory nerves. The qualities of experience, therefore, are somehow associated with the properties of sensory nerves. It was Johannes Müller who first stated this proposition in scientific form, and his statement has become known as the 'doctrine of specific nerve energies'.

Müller's doctrine of specific nerve energies

Müller's monumental contribution (1842) to our understanding of sensory processes lies in his formal statement that the brain receives information about external objects only by way of the sensory nerves. Activity in nerves, then, represents coded or symbolic data concerning the stimulus object. It is essential to note that Müller recognized only the five classical senses, the sense of touch incorporating for him all the qualities of experience that we derive from stimulation of the body. He wrote:

Sensation is a property common to all the senses; but the kind of sensation is different in each: thus, we have the sensations of light, of sound, of taste, of smell, and of feeling or touch. By feeling and touch we understand the peculiar kind of sensation of which the ordinary sensitive nerves generally, as the trigeminal, vagus, glossopharyngeal, and spinal nerves, are susceptible; the sensations of itching, or pleasure and pain, of heat and cold, and those excited by the act of touch in its more limited sense, are varieties of this mode of sensation.

For Müller, then, the somaesthetic sensations are a function of a unitary sensory system. The various qualities of somatic sensory experience provide no difficulty for the theory; no more, that is, than the different qualities of form, depth and colour perception provide difficulty for anyone regarding the visual system as a single integrated system.

Müller was uncertain, at that time, whether the quality of sensation is due to some specific energy inherent in each of the sensory nerves themselves, or whether it is due to some special properties of the brain areas at which the nerves terminate. By the late nineteenth and early twentieth centuries, however, it was apparent that nerve impulses are essentially the same in all sensory nerves, and it was concluded that the quality of sensation is given by the termination of the nerves in the brain. The impact of all this was a search for a terminal centre in the brain for each of the sensory nerves.

Müller's concept, then, was that of a straight-through system from the sensory organ to the brain centre responsible for the sensation. Since the cortex is seemingly at the 'top' of the nervous system, a search was made for cortical centres. Visual and auditory projections to cortex were found very early, and it was assumed that these cortical areas were the seat of seeing and hearing. The physiologists of the day were so convinced of the truth of this doctrine that DuBois-Reymond (see Boring, 1942) proposed that if the auditory nerve could be connected to the visual cortex, and the visual nerve to the auditory cortex, then we would see thunder and hear lightning!

We now know that we can hear quite well without auditory cortex. It is not essential for the perception of pure auditory tones; rather, its primary role is to analyse complex, sequential auditory information (Neff, 1961). It is now also apparent that the role of the visual cortex in seeing is far more complex than it seemed at first. Other areas, such as nuclei in the brainstem, play an important role in vision, and higher primates can see boundaries and contours in the absence of visual cortex (Weiskrantz, 1963). None of this was known in the late nineteenth century, however, and it was assumed that each sensory nerve must have a distinct brain centre. It was at this time that Max von Frey, a physician, first began to contemplate these problems and between 1894 and 1895 he published a series of articles in which he proposed a theory of the cutaneous senses. This theory was expanded during the next fifty years, and is the basis of modern-day specificity theory.

Von Frey's theory

The way von Frey developed his theory (Boring, 1942) makes a fascinating story in the history of science. He had three kinds of information that he put together to form it. The first was Müller's doctrine of specific nerve energies. It was apparent to him, as it was to others, that Müller's notion of a single sense of touch or 'feeling' was inadequate. The great physicist and physiologist Helmholtz proposed that there must be thousands of different specific auditory fibres, one kind for each discriminably different sound. Volkmann had similarly proposed that there must be several kinds of specific nerve fibres from the skin, one for each quality of cutaneous sensation. It was reasonable, then, for von Frey to expand Müller's concept to four major cutaneous modalities: touch, warmth, cold, and pain, each presumably with its own special projection system to a brain centre responsible for the appropriate sensation.

The second kind of information von Frey had was the spot-like distribution of warmth and cold sensitivity at the skin. He made two simple devices that are still used in neurological tests. He put a pin on a spring, and could gauge the pressure on the pin necessary to produce prick-pain, thus finding pain spots. He also put two-inch snippets of horse-tail hairs on pieces of wood and made 'von Frey hairs' to map out distributions of touch spots. Thus he believed that the skin comprises a mosaic of four types of sensory spots: touch, cold, warmth, and pain.

The third kind of information used by von Frey derived from the development, during the nineteenth century, of chemical techniques to study the fine structure of body tissues. Anatomists used particular chemicals to stain thin slices of tissue from all parts of the body, and then observed the tissues through a microscope. When they examined the skin in this way, they found a variety of specialized structures. To achieve immortality of sorts, some of the anatomists named the specialized structures after themselves. Thus, we still know these structures as Meissner corpuscles, Ruffini end-organs, Krause end-bulbs, Pacinian corpuscles and so forth. Two types of structure were so common that no one dared attach his name to them: the free nerve-endings that branch out into the upper layers of the skin, and the nerve fibres that are wrapped around hair follicles.

The way von Frey utilized these three kinds of information is a remarkable example of scientific deduction. He reasoned as follows: since the free nerve-endings are the most commonly found, and pain spots are found almost everywhere, the free nerve-endings are pain receptors. Furthermore, since Meissner corpuscles are frequently found at the fingers and palm of the hand where touch spots are most abundant and most sensitive, they (in addition to the fibres surrounding hair follicles) are the touch receptors. The next association was an imaginative deduction: he noted that the conjunctivum of the eye and the tip of the penis are both sensitive to cold, but the conjunctivum is not sensitive to warmth and the penis is not sensitive to pressure; moreover, Krause end-bulbs are found in both locations; therefore, he concluded, Krause end-bulbs are cold receptors. Finally, he had one major sensation – warmth – left over, and one major receptor – Ruffini end-organs – so he proposed that Ruffini end-organs are warmth receptors.

Von Frey's theory dealt only with receptors. Others carried on, however, and sought specific fibres from the receptors to the spinal cord, then specific pathways in the spinal cord itself.

Extensions of von Frey's theory. Following von Frey's postulation of four modalities of cutaneous sensation, each having its own type of specific nerve-ending, the separation of modality was extended to peripheral nerve fibres (see chapter 4). Ingenious experiments (reviewed by Bishop, 1946; Rose and Mountcastle, 1959; Sinclair, 1967) were carried out to show that there is a one-to-one relationship between receptor type, fibre size, and quality of experience. The concept of modality separation in peripheral nerve fibres represents the most literal interpretation of Müller's doctrine of specific nerve energies. Since fibre-diameter groups are held to be modality specific, the theory imparts 'specific nerve energy' on the basis of fibre size, so that specificity theorists speak of A-delta-fibre pain and C-fibre pain, of touch fibres and cold fibres as though each fibre group had a straight-through transmission path to a specific brain centre.

Finally, a search was made for the 'pain pathway' in the spinal cord (Keele, 1957). Experiments using animal subjects suggested that the anterolateral quadrant of the spinal cord was critically important for pain sensation. This was further reinforced by the observation by Spiller that a man who suffered damage in this area of his spinal cord was analgesic (felt no pain) in part of his body below the level of the lesion. Spiller subsequently induced a neurosurgeon named Martin to cut the anterolateral spinal cord in patients who suffered pain. The success of the operation led to the widespread performance of anterolateral cordotomy (usually known simply as 'cordotomy') for the relief of pain. As a consequence, the spinothalamic tract which ascends in the anterolateral cord has come to be known as 'the pain pathway'.

The location of the 'pain centre' is still a source of debate among specificity theorists. Head (1920) proposed that it is located in the thalamus because cortical lesions or excisions rarely abolish pain. Indeed, they may make it worse. Thus, the thalamus is held by some to contain the pain centre, and the cortex is assumed to exert inhibitory control over it.

Analysis of specificity theory

Von Frey's designation of the free nerve-endings as pain receptors is the basis of specificity theory. Its solution to the puzzle of pain is simple: specific pain receptors in body tissue project via pain fibres and a pain pathway to a pain centre in the brain. Despite its apparent simplicity, the theory has three facets, each representing a major assumption. The first of these, that receptors are specialized, is physiological in nature and has achieved the proportions of a genuine biological law. The remaining two assumptions, anatomical and psychological in nature, are not supported by the facts.

The physiological assumption. Von Frey's assumption that skin receptors are differentiated to respond to particular stimulus dimensions represents a major extension of Müller's concept of the 'specific irritability' of receptors. The assumption is that each of the four types of receptors has one form of energy to which it is especially sensitive. This concept of physiological specialization of skin receptors provides the power of von Frey's theory and appears to be the main reason for its survival through the decades. Sherrington (1900, p. 995) stated it in a manner that is acceptable to all students of sensory processes:

The sensorial end-organ is an apparatus by which an afferent nerve fibre is rendered distinctively amenable to some particular physical agent, and at the same time rendered less amenable to, i.e. is shielded from, other excitants. It lowers the value of the limen of one particular kind of stimulus, it heightens the value of the limen of stimuli of other kinds.

The beauty of Sherrington's definition of receptor specificity in terms of the lowest limen (or threshold) for a particular stimulus is that it makes no assumptions concerning the eventual psychological experience. This concept of the 'adequate stimulus' (Sherrington, 1906) is so generally accepted that it is rightfully considered to be a biological principle or law.

The anatomical assumption. It is von Frey's anatomical assumption that is the most specific, the most obviously incorrect and the least relevant aspect of the theory. Von Frey assumed that a single morphologically specific receptor lay beneath each sensory spot on the skin and he assigned a definite receptor type to each of the four modalities. He postulated his correlations on the basis of logical deduction rather than experimental evidence and he was fully aware of their defects and weaknesses. Regarding the correlation between cold and Krause's end-organs, von Frey (1895; see Dallenbach, 1927) notes that 'out of this supposition arises an obviously serious difficulty, because the ability to feel cold appears not only at the spots but also at the surrounding skin.' He also acknowledges that 'whether the number of end organs in the skin is sufficient to account for all the cold spots is a question that has yet to be decided'.

The crucial experiment of making a histological examination of the skin under carefully mapped temperature spots has been performed at least a dozen times (see Melzack and Wall, 1962), without a single investigator finding any support for von Frey's anatomical correlations. Indeed Donaldson's (1885) and Goldscheider's (1886) earlier demonstrations of only free nerve-endings beneath temperature spots have been confirmed repeatedly. Even Ruffini (1905), from the viewpoint of the histologist, noted that there are not four but a numberless variety of receptor types and considered the correlations to be nonsense. Weddell and Sinclair, during the 1950s, again destroyed every conceivable aspect of von Frey's anatomical assumption. Yet the theory has held its ground without a single counterattack. The reason, it appears, is that the anatomical assumption usually attacked lies at the periphery of von Frey's concept. The assumption that skin receptors have specialized physiological properties remains valid regardless of the correctness or incorrectness of the particular anatomical correlations suggested by von Frey. Free nerve-endings may all look alike yet each may have highly specialized properties (see chapter 4).

The psychological assumption. It is the assumption that each psychological dimension of somaesthetic experience bears a one-to-one relation to a single stimulus dimension and to a given type of skin receptor that is the most questionable part of von Frey's theory (Melzack and Wall, 1962). Like all psychological theories, von Frey's theory has an implicit conceptual nervous system; and the model is that of a fixed, direct-line communication system from the skin to the brain – of distinct nerves and pathways of four different qualities (analogous to the differently coloured wires of an electrical circuit) running from four specific kinds of stimulus transducers in the skin to four specific receivers in the brain. Figure 23 illustrates the modern-day specificity conception of the pain projection system. Despite its obvious sophistication, showing free nerve-endings, anterolateral pathway, and so forth, it is essentially similar to Descartes' concept of pain (Figure 22, p. 127) which was proposed 300 years earlier. It depicts a fixed, straight-through conceptual nervous system. It is precisely this facet of the specificity concept, which imputes a direct, invariant relationship between a psychological sensory dimension and a physical stimulus dimension, that has led to attempts at repudiation of the doctrine of specificity in its entirety. Melzack and Wall (1965, p. 971) have analysed von Frey's psychological assumption:

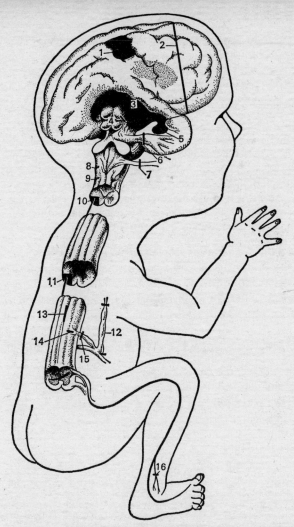

Figure 23 MacCarty and Drake's (1956, p. 208) schematic diagram illustrating various surgical procedures designed to alleviate pain: 1, gyrectomy; 2, prefrontal lobotomy; 3, thalamotomy; 4, mesencephalic tractotomy; 5, hypophysectomy; 6, fifth-nerve rhizotomy; 7, ninth-nerve neurectomy; 8, medullary tractotomy; 9, trigeminal tractotomy; 10, cervical cordotomy; 11, thoracic cordotomy; 12, sympathectomy; 13, myelotomy; 14, Lissauer tractotomy; 15, posterior rhizotomy; 16, neurectomy.

Consider the proposition that the skin contains 'pain receptors'. To say that a receptor responds only to intense, noxious stimulation of the skin is a physiological statement of fact; it says that the receptor is specialized to respond to a particular kind of stimulus. To call a receptor a 'pain receptor', however, is a psychological assumption: it implies a direct connection from the receptor to a brain centre where pain is felt, so that stimulation of the receptor must always elicit pain and only the sensation of pain. It further implies that the abstraction or selection of information concerning the stimulus occurs entirely at the receptor level and that this information is transmitted faithfully to the brain. The crux of the revolt against specificity, then, is against psychological specificity. This distinction between physiological specialization and psychological assumption also applies to peripheral fibres and central projection systems.

The facts of physiological specialization provide the power of specificity theory. Its psychological assumption is its weakness. This assumption will now be examined in the light of the psychological, clinical, and physiological evidence concerning pain (Melzack and Wall, 1962, 1965).

Psychological evidence. The psychological evidence on pain described in chapter 2 fails to support the assumption of a one-to-one relationship between pain perception and intensity of the stimulus. Instead, the evidence suggests that the amount and quality of perceived pain are determined by many psychological variables in addition to the sensory input. For example, American soldiers wounded at the Anzio beach-head 'entirely denied pain from their extensive wounds or had so little that they did not want any medication to relieve it' (Beecher, 1959, p. 165), presumably because they were overjoyed at having escaped alive from the battlefield. If the men had felt pain, even pain sensation devoid of negative affect, they would, it is reasonable to assume, have reported it, just as lobotomized patients report that they still have pain but it does not bother them. Instead, these men 'entirely denied pain'. Similarly, Pavlov's dogs that received electric shocks, burns, or cuts, followed consistently by the presentation of food, eventually responded to these stimuli as signals for food and failed to show 'even the tiniest and most subtle' (Pavlov, 1927, p. 30) signs of pain. If these dogs felt pain sensation, then it must have been nonpainful pain (Nafe, 1934) or the dogs were out to fool Pavlov and simply refused to reveal that they were feeling pain. Both possibilities, of course, are absurd. The inescapable conclusion from these observations is that intense noxious stimulation can be prevented from producing pain, or may be modified to provide the signal for eating behaviour.

The concept of four rigid modalities of cutaneous experience has been criticized by Head (1920), Nafe (1934), Livingston (1943), Hebb (1949), Weddell (1955), Sinclair (1955) and many others. The dimensions of somaesthetic perceptions have never been experimentally determined, so that psychologists are unable to agree on the number of distinctly different sensory qualities produced by skin stimulation (Titchener, 1920). The number chosen by von Frey is based purely on conjecture. To say, for example, that itch is produced by a particular pattern of stimulation of 'pain receptors' (Bishop, 1946) indicates only that the patterning of the input is more important in determining the different qualities of experience than any modality label we might arbitrarily attach to the receptor (Nafe, 1934). The four modalities of von Frey represent broad categories of different perceptual experiences that have been labelled in terms of those perceptions which are most easily named.

Clinical evidence. Phantom limb pain, causalgia, and the neuralgias provide a dramatic refutation of the concept of a fixed, direct-line nervous system. We have already noted in chapter 3 that:

1. Surgical lesions of the peripheral and central nervous system have been singularly unsuccessful in abolishing these pains permanently.

2. Gentle touch, vibration, and other non-noxious stimuli can trigger excruciating pain, and sometimes pain occurs spontaneously for long periods without any apparent stimulus.

3. The pains and new 'trigger zones' may spread unpredictably to unrelated parts of the body where no pathology exists.

4. Pain from hyperalgesic skin areas often occurs after long delays and continues long after removal of the stimulus, which implies a remarkable temporal and spatial summation of inputs in the production of these pain states.

These clinical facts defy explanation in terms of a rigid, straight-through specific pain system.

Physiological evidence. There is convincing physiological evidence (see chapter 4) that specialization exists within the somaesthetic system, but none to show that stimulation of one type of receptor, fibre, or spinal pathway elicits sensations in only a single psychological modality. Specialized fibres exist that respond only to intense stimulation, but this does not mean that they are 'pain fibres' – that they must always produce pain, and only pain, when they are stimulated. Similarly, central cells that respond exclusively or maximally to noxious stimuli are not 'pain cells'. There is no evidence to suggest that they are more important for pain perception and response than all the remaining somaesthetic cells that signal characteristic firing patterns about multiple properties of the stimulus, including noxious intensity. The view that only the cells that respond exclusively to noxious stimuli subserve pain and that the outputs of all other cells are no more than background noise is purely a psychological assumption and has no physiological basis. Physiological specialization is a fact that can be retained without acceptance of the psychological assumption that pain is determined entirely by impulses in a straight-through transmission system from the skin to a pain centre in the brain. Melzack and Wall (1962, p. 349) have noted that:

In Müller's formulation of the doctrine of specific nerve energies, the varieties of cutaneous experience are subserved by a single integrated system. Von Frey's postulation of four modalities is able to account for at least four different qualities of somaesthetic experience, but it necessarily divides the somaesthetic system into four separate subsystems, each having a direct-line transmission route to a specific termination in the brain. Müller's integrated, unspecialized conceptual nervous system was thus replaced by four specialized, unintegrated subsystems.

We can have the advantages of an integrated somaesthetic system comprised of specialized component parts if we relinquish the concept that the various sensory qualities are determined by the terminations of the ascending fibres in the brain .... There is no evidence for discrete thalamic or cortical 'centres' for any dimensions of somaesthetic experience. We must assume, therefore, that the dimensions of somaesthetic perception are subserved by different patterns of excitation evoked in the brain by different sensory stimuli. The arrival of sensory messages at the thalamus and cortex appears to mark only the beginning of a web of activity travelling in all directions to widespread areas of the central nervous system. ... In view of all this the idea of 'terminations' in the brain becomes a difficult concept: where does a pattern of excitation terminate? Surely not in the thalamus and cortex which appear to act as the hub of changing, constantly on-going processes.

If somaesthetic sensory qualities are not determined by specific terminal centres but by particular nerve impulse patterns coursing through widespread portions of the brain, there is no longer the theoretical necessity for postulating a separate system for each sensory quality. Pathways do indeed diverge, and there is no question of the specialization of central pathways for particular functions. But if we maintain the distinction between physiological specialization for information transmission and the perception and response that eventually occur, we are left free to take cognizance of any degree of specialization the data call for without encountering the difficulties of separate modality transmission routes.

Pattern theory

As a reaction against the psychological assumption in specificity theory, other theories have been proposed which can be grouped under the general heading of 'pattern theory'. Goldscheider (1894), initially one of the champions of von Frey's theory, was the first to propose that stimulus intensity and central summation are the critical determinants of pain.

Goldscheider was profoundly influenced by studies of pathological pain, especially those by Naunyn (1889) on *tabes dorsalis*, which occurs in patients suffering the late stages of syphilis. (It is rarely seen in our present age of 'miracle drugs'.) *Tabes* is characterized by degeneration in the dorsal spinal cord and dorsal roots, and one of its major symptoms is the temporal and spatial summation of somatic input in producing pain (Noordenbos, 1959). Successive, brief applications of a warm test tube to the skin of a tabetic patient are at first felt only as warm, but then feel increasingly hot until the patient cries out in pain as though his skin is being burned. Such summation never occurs in the normal person, who simply reports successive applications of warmth. Similarly, a single pin prick, which produces a momentary, sharp pain in normal subjects, evokes a diffuse, prolonged, burning pain in tabetic patients.

Not only are the intensity and duration of pain out of proportion to the stimulus, but there is often a remarkable delay in the onset of pain. A pin prick may not be felt until many seconds later – usually a few seconds but sometimes as long as forty-five seconds (Noordenbos, 1959). This astonishingly long delay was utilized dramatically by some professors of neurology (when *tabes* was a relatively common disease) who wanted this feature to be remembered by their students. At the end of a clinical demonstration period, they pin-pricked the tabetic patient and then proceeded to take off their white coats, put on their suit jackets, announce the topic and date of the next class, walk out of the door and close it – and at that point the patient yelled out in pain. Observations such as these had a powerful impact on Goldscheider, who was compelled to conclude that mechanisms of central summation, probably in the dorsal horns of the spinal cord, were essential for any understanding of pain mechanisms.

Goldscheider's pattern, or summation, theory proposes that the particular patterns of nerve impulses that evoke pain are produced by the summation of the skin sensory input at the dorsal horn cells. According to this concept, pain results when the total output of the cells exceeds a critical level as a result of either excessive stimulation of receptors that are normally fired by non-noxious thermal or tactile stimuli, or pathological conditions that enhance the summation of impulses produced by normally non-noxious stimuli. The long delays and persistent pain observed in pathological pain states, Goldscheider assumed, are due to abnormally long time-periods of summation. He proposed, moreover, that the spinal 'summation path' that transmits the pain signals to the brain consists of slowly conducting, multi-synaptic fibre chains. The large fibres that project up the dorsal column pathways were presumed to carry specific information about the tactile discriminative properties of cutaneous sensation.

Several theories have emerged from Goldscheider's concept. All of them recognize the concept of patterning of the input, which is essential for any adequate theory of pain. But some ignore the facts of physiological specialization, while others utilize them in proposing mechanisms of central summation.

Peripheral pattern theory

The simplest form of pattern theory deals primarily with peripheral rather than central patterning. That is, pain is considered to be due to excessive peripheral stimulation that produces a pattern of nerve impulses which is interpreted centrally as pain. The pattern theory of Weddell (1955) and Sinclair (1955) is based on the earlier suggestion by Nafe (1934) that all cutaneous qualities are produced by spatial and temporal patterns of nerve impulses rather than by separate modality-specific transmission routes. The theory proposes that all fibre endings (apart from those that innervate hair cells) are alike, so that the pattern for pain is produced by intense stimulation of nonspecific receptors.

The physiological evidence, however, reveals a high degree of receptor-fibre specialization. The pattern theory proposed by Weddell and Sinclair, then, fails as a satisfactory theory of pain because it ignores the facts of physiological specialization. We have already noted in chapter 4 that it is more reasonable to assume that the specialized physiological properties of each receptor-fibre unit (such as thresholds to different stimuli, adaptation rates, and size of receptive field) play an important role in determining the characteristics of the temporal patterns that are generated when a stimulus is applied to the skin.

Central summation theory

The analysis of phantom limb pain, causalgia and the neuralgias in chapter 3 indicates that part, at least, of their undelying mechanisms must be sought in the central nervous system. Livingston (1943) was the first to suggest specific central neural mechanisms to account for the remarkable summation phenomena in these pain syndromes. He proposed that pathological stimulation of sensory nerves (such as occurs after peripheral nerve damage) initiates activity in reverberatory circuits in neuron pools in the spinal cord (see Figure 20, p. 118). This abnormal activity can then be triggered by normally non-noxious inputs and generate volleys of nerve impulses that are interpreted centrally as pain.

Livingston's theory is especially powerful in explaining phantom limb pain. He proposed that the initial damage to the limb, or the trauma associated with its removal initiates abnormal firing patterns in closed, self-exciting neuron loops in the dorsal horns of the spinal cord, which send volleys of nerve impulses to the brain that give rise to pain. Moreover, the reverberatory activity may spread to adjacent neurons in the lateral and ventral horns and produce autonomic and muscular manifestations in the limb, such as sweating and jerking movements of the stump. These, in turn, produce further sensory input, creating a 'vicious circle' between central and peripheral processes that maintains the abnormal spinal cord activity (Figure 24). Even minor irritations of the skin or nerves near the site of the operation can then feed into these active pools of neurons and keep them in an abnormal, disturbed state over periods of years. Impulse patterns that would normally be interpreted as touch may now trigger these neuron pools into greater activity, thereby sending volleys of impulses to the brain to produce pain. In addition, emotional disturbance may evoke neural activity that feeds into the abnormal neuron pools. Once the abnormal cord activity has become self-sustaining, surgical removal of the peripheral sources of input may not stop it. Rather, clinical procedures that modulate the sensory input, such as local anaesthetic

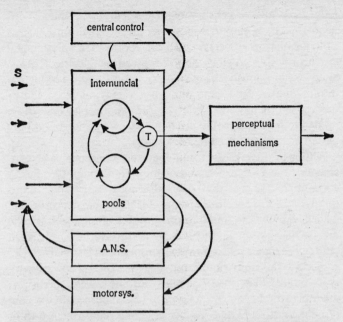

Figure 24 Schematic diagram of W. K. Livingston's (1943) theory of pathological pain states. The intense stimulation (S) resulting from nerve and tissue damage activates fibres that project to internuncial neuron pools in the spinal cord, creating abnormal reverberatory activity in closed self-exciting neuron loops. This prolonged, abnormal activity bombards the spinal cord transmission (T) cells which project to brain mechanisms that underlie pain perception. The abnormal internuncial activity also spreads to lateral and ventral horn cells in the spinal cord, activating the autonomic nervous system (A.N.S.) and motor system, producing sweating, jactitations, and other manifestations. These, in turn, produce further abnormal input, thereby creating a 'vicious circle'. Brain activities such as fear and anxiety evoked by pain also feed into and maintain the abnormal internuncial pool activity.

injections or physiotherapy, may again reinstate normal cord activity.

Gerard (1951) has suggested a theory that is similar in concept, although different in hypothetical mechanism. He proposed that a peripheral nerve lesion may bring about a temporary loss of sensory control of firing in spinal cord neurons. These may then begin to fire in synchrony, just as isolated bits of nerve tissue in an appropriate solution fire synchronously, possibly due to spread of electrical fields. Such synchronously firing neuron pools 'could recruit additional units, could move along in the grey matter, could be maintained by impulses different from and feebler than those needed to initiate it, could discharge excessive and abnormally patterned volleys to the higher centres'.

Although Livingston's and Gerard's concepts have considerable power in explaining phantom limb pain, they fail to account for the fact that surgical lesions of the spinal cord often do not abolish the pain. Instead, the neurosurgical evidence points to mechanisms in the brain itself. If the crucial mechanism lay in the spinal cord dorsal horns, then cutting the major sensory routes through which spinal activity projects to the brain should stop the pain. Yet it is now generally recognized that, once the pain syndrome is well established, attempts to relieve it by surgery of spinal cord pathways are often ineffective. White and Sweet (1969) report the return of phantom limb pain after cordotomy in seven out of eighteen lower limb and three out of four upper limb cases. Even bilateral cordotomy may fail. Efforts have been made to find 'the leak' in the pain projection system, and the multi-synaptic propriospinal fibre chain has been proposed as one possibility (Noordenbos, 1959). But if there is a leak, and impulses ascending the cord determine pain without further elaboration, it is hard to imagine that so small an input (after the extensive surgical cuts) can produce such massive pain. Surgeons have therefore turned to the dorsal columns, the traditional 'touch-proprioception pathway' in the spinal cord, particularly for cramping pains in the phantom limb. Yet here too the operation is usually ineffective in producing a permanent cure (White and Sweet, 1969).

This emphasis on spinal cord activity is avoided by Hebb (1949), who suggests that synchronized firing in thalamo-cortical neural circuits provides the signal for pain. Hebb is particularly explicit in his concept that pain is determined by central summation mechanisms. He notes that pain frequently occurs after lesions at any level of the somaesthetic pathway. The loss of sensory control of patterned thalamo-cortical activities, he proposes, would produce excessive, synchronous firing in brain cells, which would disrupt the patterned activities that normally subserve perceptual and cognitive processes. The disruption itself, he suggests, *is* pain.

Sensory interaction theory

Related to theories of central summation is the theory that a specialized input-controlling system normally prevents summation from occurring, and that destruction of this system leads to pathological pain states. This theory derives from Goldscheider's original concept, and proposes the existence of a rapidly conducting fibre system which inhibits synaptic transmission in a more slowly conducting system that carries the signals for pain. Historically (see Melzack and Wall, 1965), these two systems are identified as the epicritic and protopathic (Head, 1920), fast and slow (Bishop, 1946), phylogenetically new and old (Bishop, 1959), and myelinated and unmyelinated (Noordenbos, 1959) fibre systems. Under pathological conditions, the fast system loses its dominance over the slow one, and the result is protopathic sensation (Head, 1920), slow pain (Bishop, 1946), diffuse burning pain (Bishop, 1959), or hyperalgesia (Noordenbos, 1959).

Noordenbos' theory (Figure 25) represents an especially important contribution to sensory-interaction concepts. The small fibres are conceived as carrying the nerve impulse patterns that produce pain, while the large fibres inhibit transmission. A shift in the ratio of large-to-small fibres in favour of the small fibres would result in increased neural transmission, summation, and excessive pathological pain. Just as important as the input control by the large fibres, in Noordenbos' concept, is the idea of a multi-synaptic afferent system in the spinal cord. It stands in marked contrast to the idea of a straight-through system, and has the power to explain why spinothalamic cordotomy may fail to abolish pain. The diffuse, extensive connections within the ascending multi-

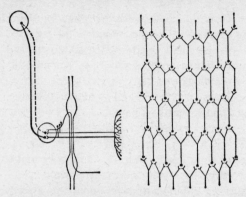

Figure 25 Noordenbos' (1959) concept of pain mechanisms. *Left:* small-diameter, slowly conducting somatic afferents and small visceral afferents (which travel through the sympathetic ganglia) project onto cells in the dorsal horn of the spinal cord. The summation of inputs from the small fibres produces the neural patterns that are transmitted to the brain to produce pain. The large-diameter fibres inhibit transmission of impulses from the small fibres and prevent summation from occurring. A selective loss of large fibres brings about a loss of inhibition and thereby increases the probability of summation and abnormal pain phenomena. The small fibre projection system – drawn as a dashed line – is indicated to be multi-synaptic. *Right:* Noordenbos' representation of the 'Multi-synaptic Afferent System' in the spinal cord. The diffuse, widespread conduction through the system, Noordenbos suggests, is the basis of 'the leak' of nerve signals that evoke pain even after extensive surgical section of the anterolateral pathways.

synaptic afferent system, he proposes, can rarely (if ever) be totally abolished (unless the whole spinal cord is cut), so that there is always a 'leak' for impulses to ascend to the brain to produce pain.

Noordenbos' theory has considerable power in explaining many of the pathological pain states described in chapter 3. The concept of a shift in the fibre-diameter groups in favour of the small fibres is consistent with the observed relative loss of large fibres after peripheral nerve injury, and is able to explain the delays, temporal and spatial summation, and many other properties of pathological pain. The development of 'girdle pains' after cordotomy, he proposes, may similarly be due to the relative loss of the large fibres in the anterolateral pathways which still leaves the small diffusely conducting fibres untouched. Sympathectomy, in contrast, would tend to destroy the small fibres, leaving a predominance of inhibitory large fibres which would decrease the tendency to summation and, consequently, the level of pain. Noordenbos' theory, like Livingston's, represents a major theoretical advance towards an understanding of the puzzle of pain.

Affect theory of pain

The theory that pain is a sensory modality is relatively recent. A much older theory, dating back to Aristotle, considers pain to be an emotion – the opposite of pleasure – rather than a sensation. Indeed, this idea of pain is part of an intriguing and usually neglected bit of history (Dallenbach, 1939). At the turn of the century, a bitter battle was fought on the question of pain specificity. Von Frey argued that there are specific pain receptors, while Goldscheider contended that pain is produced by excessive skin stimulation and central summation. But there was a third man in the battle – H. R. Marshall (1894), a philosopher and psychologist – who said, essentially, 'a plague on both your houses; pain is an emotional quality, or *quale*, that colours all sensory events'. He admitted the existence of a pricking-cutting sense, but thought that pain was distinctly different. All sensory inputs, as well as thoughts, could have a painful dimension to them, and he talked of the pain of bereavement, the pain of listening to badly played music. His extreme approach was, of course, open to criticism. Sherrington (1900), for example, noted that the pain of a scalded hand is different from the 'pain' evoked in a musicologist by even the most horrible discord. Marshall was soon pushed off the field. But if a less extreme view is taken of his concept, it suggests an important yet neglected dimension of pain. For pain does not have just a sensory quality; it also has a strong negative affective quality that drives us into activity (Figure 26). We are compelled to do something about it, to take the most effective course of action to stop it, and,

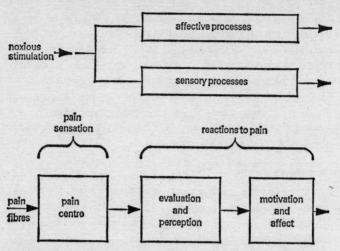

Figure 26 *Top:* diagram of Marshall's (1894) concept of pain as an affective quality or *quale.* Intense stimulation of the skin activates two parallel systems: one is the basis of the affective properties of the experience, the other underlies the sensory properties. *Bottom:* diagram of the concept, implicit in specificity theory, that motivation and affect are reactions to pain, but are not part of the primary pain sensation. (from Melzack and Casey, 1968)

of course, this behaviour is in the realm of emotion and motivation.

That pain is comprised of both sensory and affective dimensions was clear to Sherrington (1900) who proposed simply that 'mind rarely, probably never, preceives any object with absolute indifference, that is, without "feeling" ... affective tone is an attribute of all sensation, and among the attribute tones of skin sensation is skin pain.' Introspectionist psychologists at the turn of the century also made a sharp distinction between the sensory and the affective qualities of pain. Titchener (1909) was convinced that there is a continuum of feeling in conscious experience, distinctly different from sensation, that ranges through all the degrees of pleasantness and unpleasantness. 'The pain of a toothache,' he wrote, 'is localized at a particular place, "in the tooth"; but the unpleasantness of it suffuses the whole of present experience, is as wide as consciousness. The word "pain" ... often means the whole toothache experience.'

The remarkable development of sensory physiology and psychophysics during the twentieth century has given momentum to the concept of pain as a sensation and has overshadowed the role of affective and motivational processes. The sensory approach to pain, however, valuable as it has been, fails to provide a complete picture of pain processes. The neglect of the motivational features of pain underscores a serious schism in pain research. Characteristically, textbooks in psychology and physiology deal with 'pain sensation' in one section and 'aversive drives and punishment' in another, with no indication that both are facets of the same phenomenon. This separation reflects the widespread acceptance of von Frey's specificity theory of pain, with its implicit psychological assumption that 'pain impulses' are transmitted from specific pain receptors in the skin directly to a pain centre in the brain.

The assumption that pain is a primary sensation has relegated motivational (and cognitive) processes to the role of 'reactions to pain' (Figure 26), and has made them only 'secondary considerations' in the whole pain process (Sweet, 1959). It is apparent, however, that sensory, motivational, and cognitive processes occur in parallel, interacting systems at the same time. As we noted in chapter 4, motivational-affective processes must be included in any satisfactory theory of pain.

Evaluation of the theories

When we consider all the theories examined so far, we see that the 'specific-modality' and 'pattern' concepts of pain, although they appear to be mutually exclusive, both contain valuable concepts that supplement one another. Recognition of receptor specialization for the transduction of particular kinds and ranges of cutaneous stimulation does not preclude acceptance of the concept that the information generated by skin receptors is coded in the form of patterns of nerve impulses. The law of the adequate stimulus can be retained without also accepting a narrow, fixed relationship between receptor specialization and perceptual experience.

It is clear that von Frey made an important contribution that must be retained in any theoretical formulation. He proposed that the receptors of the skin are not all alike but are differentiated with respect to lowest threshold to particular energy categories. This concept of receptor specialization continues to play a salient role in sensory physiology and psychology. Indeed, the recent evidence indicates a greater degree of receptor specialization than von Frey himself could ever have foreseen. The theory, however, encounters serious difficulties. It implies a narrow one-to-one relationship between psychological sensory dimensions and physical stimulus dimensions that is inadmissible in view of our current knowledge about pain.

Similarly, there can no longer be any doubt that temporal and spatial patterns of nerve impulses provide the basis of our sensory perceptions. The coding of information in the form of nerve impulse patterns is a fundamental concept in contemporary neurophysiology and psychology. Yet the peripheral pattern theory formulated by Weddell and Sinclair fails to provide an adequate account of pain mechanisms. It does not recognize the facts of physiological specialization. It does not specify the kinds of patterns that might be related to pain. It provides no hypothesis to account for the detection of patterns by central cells. Thus, because of its vagueness, the theory falls short of being a satisfactory formulation of pain phenomena.

In contrast, the concepts of central summation and input control have shown remarkable power in their ability to explain many of the clinical phenomena of pain. Goldscheider's emphasis on central summation mechanisms is supported by the clinical observations of extraordinary temporal and spatial summation in pathological pain syndromes. Livingston's theory of spinal reverberatory activity that persists in the absence of noxious input provides a satisfactory explanation of prolonged pain. Noordenbos' concept that large fibres inhibit activity in small fibres is supported by the evidence that pathological pain is often associated with a loss of large myelinated fibres. These theories, nevertheless,

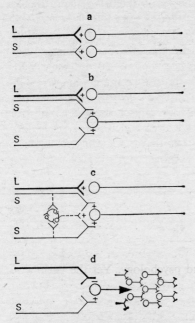

Figure 27 Schematic representation of conceptual models of pain mechanisms. A: von Frey's specificity theory. Large (L) and small (S) fibres are assumed to transmit touch and pain impulses respectively, in separate, specific, straight-through pathways to touch and pain centres in the brain. B: Goldscheider's summation theory, showing convergence of small fibres onto a dorsal horn cell. Touch is assumed to be carried by large fibres. C: Livingston's (1943) conceptual model of reverberatory circuits underlying pathological pain states. Prolonged activity in the self-exciting chain of neurons bombards the dorsal horn cell, which transmits abnormally patterned volleys of nerve impulses to the brain. D: Noordenbos' (1959) sensory interaction theory, in which large fibres inhibit (—) and small fibres excite (+) central transmission neurons. The output projects to spinal cord neurons which are conceived by Noordenbos to comprise a Multi-synaptic Afferent System.
(from Melzack and Wall, 1970, p. 3)

fail to comprise a satisfactory general theory of pain. They lack unity, and no single theory has yet been proposed that integrates the diverse theoretical mechanisms.

However, when all the theories – from specificity theory onward – are examined together (Figure 27), it is apparent that each successive theory makes an important contribution. Each provides an additional mechanism to explain some of the complex clinical syndromes or experimental data that were previously inexplicable. Despite the seemingly small differences, each change contains a major conceptual idea that has had a powerful impact on research and therapy.

6 The Gate-Control Theory of Pain

The analysis of the strengths and weaknesses of the theories of pain described in chapter 5 illuminates some of the requirements of a satisfactory new theory. Any new theory of pain, it is now apparent, must be able to account for:

1. The high degree of physiological specialization of receptor-fibre units and of pathways in the central nervous system.
2. The role of temporal and spatial patterning in the transmission of information in the nervous system.
3. The influence of psychological processes on pain perception and response.
4. The clinical phenomena of spatial and temporal summation, spread of pain, and persistence of pain after healing.

The *gate-control theory* recently proposed by Melzack and Wall (1965) has attempted to integrate these requirements into a comprehensive theory of pain. Basically, the theory proposes that a neural mechanism in the dorsal horns of the spinal cord acts like a gate which can increase or decrease the flow of nerve impulses from peripheral fibres to the central nervous system. Somatic input is therefore subjected to the modulating influence of the gate before it evokes pain perception and response. The degree to which the gate increases or decreases sensory transmission is determined by the relative activity in large-diameter (A-beta) and small-diameter (A-delta and C) fibres and by descending influences from the brain. When the amount of information that passes through the gate exceeds a critical level, it activates the neural areas responsible for pain experience and response. Like all theories, the gate-control theory has two facets: a conceptual model which is the basis of the theory, and particular explanatory mechanisms which are evoked to show how the model functions. The conceptual model will be described first, followed by the explanatory mechanisms that have been proposed.

The conceptual model

The conceptual model that underlies the gate-control theory of pain (Figure 28) is based on the following propositions:

1. The transmission of nerve impulses from afferent fibres to spinal cord transmission (T) cells is modulated by a spinal gating (SG) mechanism in the dorsal horns.

2. The spinal gating mechanism is influenced by the relative amount of activity in large-diameter (L) and small-diameter (S) fibres: activity in large fibres tends to inhibit transmission

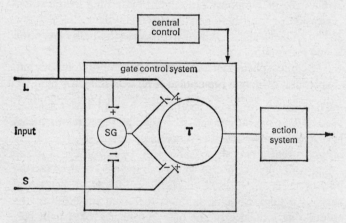

Figure 28 Schematic diagram of the gate-control theory of pain mechanisms: L, the large-diameter fibres; S, the small-diameter fibres. The fibres project to the substantia gelatinosa (SG) and first central transmission (T) cells. The inhibitory effect exerted by SG on the afferent fibre terminals is increased by activity in L fibres and decreased by activity in S fibres. The central control trigger is represented by a line running from the large fibre system to the central control mechanisms; these mechanisms, in turn, project back to the gate-control system. The T cells project to the entry cells of the action system. +, excitation; —, inhibition. (from Melzack and Wall, 1965, p. 971)

(close the gate) while small-fibre activity tends to facilitate transmission (open the gate).

3. The spinal gating mechanism is influenced by nerve impulses that descend from the brain.

4. A specialized system of large-diameter, rapidly conducting fibres (the Central Control Trigger) activates selective cognitive processes that then influence, by way of descending fibres, the modulating properties of the spinal gating mechanism.

5. When the output of the spinal cord transmission (T) cells exceeds a critical level, it activates the Action System – those neural areas that underlie the complex, sequential patterns of behaviour and experience characteristic of pain.

The small (A-delta and C) fibres, in this conceptual framework, play a highly specialized and important role in pain processes. They activate the T cells directly and contribute to their output. The activity of high-threshold small fibres, during intense stimulation, may be especially important in raising the T-cell output above the critical level necessary for pain. But the small fibres are believed (Melzack and Wall, 1965) to do much more than this. They facilitate transmission ('open the gate') and thereby provide the basis for summation, prolonged activity, and spread of pain to other body areas. This facilitatory influence provides the small fibres with greater power than any envisaged in the concept of 'pain fibres'. Yet at the same time the small-fibre impulses are susceptible to modulation by activities in the whole nervous system. This multi-faceted role of the small fibres is consistent with the psychological, clinical, and physiological evidence.

Explanatory mechanisms

Recent physiological evidence provides an explanatory basis for the conceptual model. Some of the evidence is well established. However, it has also been necessary to utilize indirect evidence in order to speculate about some aspects of the model.

Spinal gating mechanism

The substantia gelatinosa (laminae 2 and 3) appears to be the most likely site of the spinal gating mechanism (Wall, 1964; Melzack and Wall, 1965). It receives axon terminals from many of the large- and small-diameter fibres and the dendrites of cells in deeper laminae project into it (Figure 29). The substantia gelatinosa, moreover, forms a functional unit that extends the length of the spinal cord on each side. Its cells connect with one another by short fibres, and influence each other at distant sites on the same side by means of Lissauer's tract and on the opposite side by means of commissural fibres that cross the cord (Szentagothai, 1964; Wall, 1964). The substantia gelatinosa, then, consists of a highly specialized, closed system of cells throughout the length of the spinal cord on both sides; it receives afferent input from large and small fibres, and is able to influence the activity of cells that project to the brain. Melzack and Wall (1965) have proposed, therefore, that it acts as a spinal gating mechanism by modulating the conduction of nerve impulses from peripheral fibres to spinal cord transmission cells.

Figure 29 *Top:* schematic drawing of the substantia gelatinosa in relation to somatosensory fibres and dorsal horn cells that project their axons across the cord to the anterolateral pathway. Source: after Pearson, 1952, p. 515
Bottom: main components of the cutaneous afferent system in the upper dorsal horn. The large-diameter cutaneous peripheral fibres are represented by thick lines running from the dorsal root and terminating in the region of the substantia gelatinosa; one of these, as shown, sends a branch toward the brain in the dorsal column. The finer peripheral fibres are represented by dashed lines running directly into the substantia gelatinosa. The large cells, on which cutaneous afferent nerves terminate, are shown as large black spheres with their dendrites extending into the substantia gelatinosa and their axons projecting deeper into the dorsal horn. The open circles represent the cells of the substantia gelatinosa. The axons (not shown) of these cells connect them to one another and also run in the Lissauer tract (LT) to distant parts of the substantia gelatinosa.
(adapted from Wall, 1964, p. 92)

The properties of the cells in lamina 5 suggest that they are the spinal transmission (T) cells that are most likely to play a critical role in pain perception and response (Hillman and Wall, 1969). They receive inputs from the small afferent fibres from skin, viscera and muscles, and their activity is

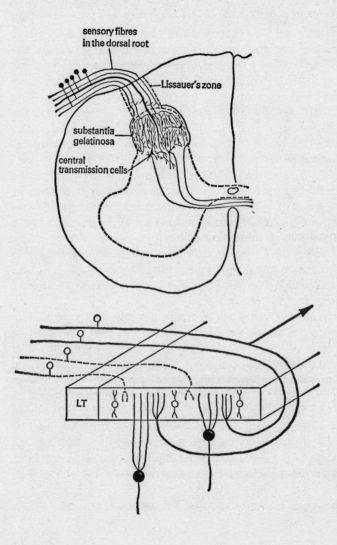

influenced by fibres that descend from the brain. They respond to a wide range of stimulus intensities, and show increasing firing rates to increasing intensities of stimulation. Furthermore, their output is influenced by the relative activity in large and small fibres.

Effects of activity in large and small fibres

Melzack and Wall (1965, 1970) propose that sensory fibres transmit patterned information, depending on the specialized properties of each receptor-fibre unit, about pressure, temperature, and chemical changes at the skin. These temporal and spatial patterns of nerve impulses have two effects at the dorsal horns: they excite the spinal cord T cells that project the information to the brain, and they activate the substantia gelatinosa which modulates or 'gates' the *amount* of information projected to the brain by the T cells. Recent studies (Hillman and Wall, 1969) have shown that activity in large fibres produces a burst of activity in lamina 5 cells followed by a period of inhibition. In contrast, activity in small fibres activates the cells and then produces prolonged activity and a facilitation of subsequent inputs (the 'wind-up' effect). The evidence, then, suggests that transmission in the dorsal horns is controlled by a spinal gating mechanism which is in turn controlled by the rival effects of large versus small afferent fibres.

There are two ways in which the cells of the substantia gelatinosa can act as a gating mechanism that influences the transmission of impulses from afferent fibre terminals to spinal cord cells (Melzack and Wall, 1970). They can act directly on the presynaptic axon terminals and thereby block the impulses in the terminals or decrease the amount of transmitter substance which they release; or they can act postsynaptically on the spinal transmission cells by increasing or decreasing their level of excitability to arriving nerve impulses. Melzack and Wall (1965) proposed that the effect is primarily presynaptic. But it is now certain (Hongo, Jankowska and Lundberg, 1968) that modulating effects are also exerted postsynaptically on the spinal transmission cells. The evidence, in other words, indicates that the presynapatic control exists but that it is coupled with a simultaneous change in the postsynaptic transmission cells.

Melzack and Wall (1965, p. 975) propose that three features of the afferent input are significant for pain: the ongoing activity which precedes the stimulus, the stimulus-evoked activity, and the relative balance of activity in large versus small fibres:

A continuous flow of nerve impulses to the spinal cord, in the absence of obvious stimulation, is carried predominantly by small myelinated and unmyelinated fibres, which tend to be tonically active and to adapt slowly, and holds the gate in a relatively open position. When a tactile, thermal or chemical stimulus is applied to the skin, many fibres start firing or increase their activity as information about the stimulus is transmitted towards the brain. Since many of the larger fibres are inactive in the absence of stimulus change, stimulation produces more activity in the large fibres than in small fibres, so that the volley fires the T cells but also partially closes the presynaptic gate. If the stimulus intensity is increased, more receptor-fibre units are recruited and the firing frequency of active units is increased (Wall, 1960). The resultant positive and negative effects of the large-fibre and small-fibre inputs tend to counteract each other, and the output of the T cells rises slowly. If stimulation is prolonged, the large fibres begin to adapt, producing a relative increase in small-fibre activity. As a result, the gate is opened further, and the output of the T cells rises more steeply. If large-fibre activity is raised at this time by vibration or scratching (which overcomes the tendency of the large fibres to adapt), the output of the cells decreases. Clearly, then, the output of the T cells may differ from the total input that converges on them depending on the balance of activity in large and small fibres.

Burgess, Petit and Warren (1968) have recently shown that the relationship between fibre diameter and adaptation rate is more complex than Melzack and Wall had assumed. They found that, in the entire receptor population innervated by *myelinated* fibres, there was no general tendency for receptors with more slowly conducting fibres to be more slowly adapting. They note, however, that within each receptor group (such as field receptors or hair receptors), 'there was a tendency for the receptors with the most rapidly conducting fibres to be the most rapidly adapting'. These observations are consistent with the assumptions of the gate theory but indicate a high degree of specialization within receptor-fibre groups and, presumably, within the cell populations influenced by these fibres in laminae 4 and 5.

The inhibitory and facilitatory effects of large and small fibres on dorsal horn cells may represent only the beginning of successive interactions between fast and slow conducting fibre systems. The dorsal column projection system and the more slowly conducting spinothalamic fibre system project to overlapping areas of the somatosensory thalamus and cortex (Noordenbos, 1959; Rose and Mountcastle, 1959). It is possible that both systems exert counteracting inhibitory and facilitatory influences, so that the modulation of the sensory input at the dorsal horns may continue in a similar way at successively higher synaptic levels.

Descending influences on the gate-control system

The evidence described in chapter 2 shows that cognitive or 'higher central nervous system processes' such as attention, anxiety, anticipation, and past experience exert a powerful influence on pain processes. It is also firmly established (chapter 4) that stimulation of the brain activates descending efferent fibres which can influence afferent conduction at the earliest synaptic levels of the somaesthetic system. Thus it is possible for brain activities subserving attention, emotion and memories of prior experience to exert control over the sensory input. This control of spinal cord transmission by the brain may be exerted through several systems.

Reticular projections. The brainstem reticular formation, particularly the midbrain reticular areas (Hagbarth and Kerr, 1954; Taub, 1964), exert a powerful inhibitory control over information projected by the spinal transmission cells. The inhibition of activity in lamina 5 cells by descending fibres from the brain (Hillman and Wall, 1969) is at least partly due to reticulo-spinal influences on the dorsal horn gating system.

This descending inhibitory projection is itself controlled by multiple influences. Somatic projections comprise the largest input to the midbrain reticular formation. There are also projections from the visual and auditory systems (Rossi and Zanchetti, 1957). In this way, somatic inputs from all parts of the body, as well as visual and auditory inputs, are able to exert a modulating influence on transmission through the dorsal horns.

Cortical projections. Fibres from the whole cortex, particularly the frontal cortex, project to the reticular formation. Cognitive processes such as past experience and attention, which are subserved at least in part by cortical neural activity, are therefore able to influence spinal activities by way of the reticulo-spinal projection system. Cognitive processes can also influence spinal gating mechanisms by means of pyramidal (or cortico-spinal) fibres, which are known to project to the dorsal horns as well as to other spinal areas. These are large, fast-conducting fibres so that cognitive processes can rapidly and directly modulate neural transmission in the dorsal horns.

Concept of a central control trigger. It is apparent that the influence of cognitive or 'central control' processes on spinal transmission are mediated, in part at least, through the gate-control system. While some central activities, such as anxiety or excitement, may open or close the gate for all inputs from any part of the body, others obviously involve selective, localized gate activity. The observations by Pavlov (1927, 1928) and Beecher (1959) described in earlier chapters suggest that signals from the body must be identified, evaluated in terms of prior experience, localized, and inhibited *before* the action system reponsible for pain perception and response is activated.

Melzack and Wall (1965) have therefore proposed that there exists in the nervous system a mechanism, which they have called the *central control trigger*, that activates the particular, selective brain processes that exert control over the sensory input (Figure 28, p. 154). They suggest that the dorsal-column–medial-lemniscal and dorso-lateral systems could fulfil the functions of the central control trigger. The dorsal column projection system in particular has grown apace with the cerebral cortex (Bishop, 1959), carries precise information about the nature and location of the stimulus, adapts quickly to give precedence to phasic stimulus changes rather than prolonged tonic activity, and conducts so rapidly that it may not only set the receptivity of cortical neurons for subsequent afferent volleys but may also act, by way of central-control efferent fibres, on the gate-control system. Part, at least, of their functions, then, could be to activate selective brain processes such as memories of prior experience and pre-set response strategies that influence information which is still arriving over slowly conducting fibres or is being transmitted up more slowly conducting pathways.

Action system

The gate-control theory (Melzack and Wall, 1965) proposes that the action system responsible for pain experience and response is triggered when the integrated firing level of the dorsal horn T cells reaches or exceeds a critical level. Melzack and Casey (1968) have noted that the output of the T cells is transmitted towards the brain primarily by fibres in the antero-lateral spinal cord and is projected into two major brain systems: via neospinothalamic fibres into the ventrobasal thalamus and somatosensory cortex, and via medially coursing fibres, that comprise a paramedial ascending system, into the reticular formation and medial intralaminar thalamus and the limbic system (Figure 13, p. 94). Stimulation at noxious intensities evokes activity in both projection systems, and discrete lesions in each may strikingly alter pain perception and response (chapter 4).

Recent behavioural and physiological studies have led Melzack and Casey (1968) to propose (Figure 30) that:

1. The selection and modulation of the sensory input through the neospinothalamic projection system provides, in part at least, the neurological basis of the sensory-discriminative dimension of pain.

2. Activation of reticular and limbic structures through the paramedial ascending system underlies the powerful motivational drive and unpleasant affect that trigger the organism into action.

3. Neocortical or higher central nervous system processes, such as evaluation of the input in terms of past experience, exert control over activity in both the discriminative and motivational systems.

It is assumed that these three categories of activity interact with one another to provide *perceptual information* regarding the location, magnitude, and spatiotemporal properties of the noxious stimulus, *motivational tendency* toward escape or attack, and *cognitive information* based on analysis of multimodal information, past experience, and probabilty of outcome of different response strategies. All three forms of activity could then influence motor mechanisms responsible for the complex pattern of overt responses that characterize pain.

There is now a convincing body of evidence that stimulation of reticular and limbic system structures produces strong aversive drive and behaviour typical of responses to naturally occurring painful stimuli. These data together with related evidence (see chapter 4) on the effects of ablation indicate that limbic structures, although they play a role in many other functions, provide a neural basis for the aversive drive and affect that comprise the motivational dimension of pain. The fact that there are inputs from other sensory systems as well as the cutaneous system indicates that these areas are not activated exclusively by noxious stimuli. Moreover, the somatic input has access to areas involved in both approach and avoidance, and stimulation of some areas can produce either kind of response. On what basis, then, are aversive rather than approach mechanisms triggered by the input?

Melzack and Casey propose that portions of the reticular and limbic systems function as a *central intensity monitor:* that their activities are determined, in part at least, by the intensity of the T-cell output (the total number of active fibres and their

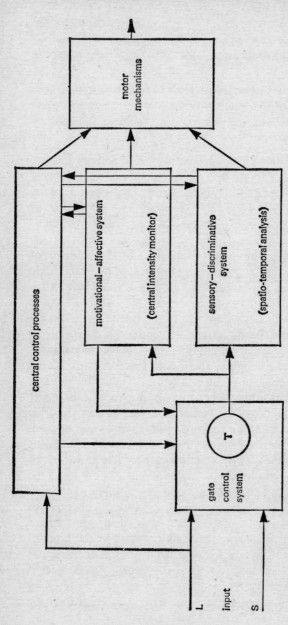

Figure 30 Conceptual model of the sensory, motivational and central control determinants of pain. The output of the T cells of the gate-control system projects to the sensory-discriminative system (via neospinothalamic fibres) and the motivational-affective system (via the paramedial ascending system). The central control trigger is represented by a line running from the large fibre system to central control processes; these, in turn, project back to the gate-control system, and to the sensory-discriminative and motivational-affective systems. All three systems interact with one another, and project to the motor system. (from Melzack and Casey, 1968)

rate of firing) after it has undergone modulation by the gate-control system in the dorsal horns. The cells in the midbrain reticular formation are capable of summation of input from spatially separate body sites (Amassian and DeVito, 1954; Bell, Sierra, Buendia and Segundo, 1964); furthermore, the post-stimulus discharge activity of some of these cells lasts for many seconds (Casey, 1966), so that their activity may provide a measure of the intensity of the total T-cell output over relatively long periods of time. Essentially, both kinds of summation transform discrete spatial and temporal information into intensity information. Melzack and Casey propose that the output of these cells, up to a critical intensity level, activates those brain areas subserving positive affect and approach tendency. Beyond that level, the output activates areas underlying negative affect and aversive drive. They suggest, therefore, that the drive mechanisms associated with pain are activated when the somatosensory input into the motivational-affective system exceeds the critical level. This notion fits well with Grastyan, Czopf, Angyan and Szabo's (1965) observations that animals seek low-intensity electrical stimulation of some limbic system structures, but avoid or actively try to stop high-intensity stimulation of the same areas. Signals from these limbic structures to motor mechanisms, together with the information derived from sensory and cognitive processes, could selectively activate neural networks that subserve adaptive response patterns.

The complex sequences of behaviour that characterize pain are determined by sensory, motivational, and cognitive processes that act on motor mechanisms. By 'motor mechanisms' (Figure 30), Melzack and Casey mean all of the brain areas that contribute to overt behavioural response patterns. These areas, we have already seen (chapter 4), extend throughout the whole of the central nervous system, and their organization must be at least as complex as that of the input systems we have primarily dealt with so far. Even 'simple' reflexes, which are generally thought to be entirely spinal in their organization, are now known to be influenced by cognitive processes: if we pick up a hot cup of tea in an expensive cup we are not likely to simply drop the cup, but jerkily put it back on the table, and *then* nurse our hand.

Central biasing mechanism

The interactions between the gate-control system and the action system are only the beginning of the complex processes that characterize pain experience and behaviour (Melzack and Wall, 1965, p. 976):

> The interactions ... may occur at successive synapses at any level of the central nervous system in the course of filtering of the sensory input. Similarly, the influence of central activities on the sensory input may take place at a series of levels. The gate-control system may be set and reset a number of times as the temporal and spatial patterning of the input is analysed and acted on by the brain.

The inhibitory influence which is exerted on spinal transmission by the midbrain reticular formation merits special examination. It provides the gate-control theory with additional power to explain some of the most puzzling phenomena of pain. We have already noted that lesions of the central tegmental tract in cats produce hyperalgesia (Melzack, Stotler, and Livingston, 1958), and electrical stimulation at points in the tract and adjacent structures (Figure 31) produces analgesia in rats (Reynolds, 1969, 1970). More recently, it has been found (Mayer, Wolfle, Akil, Carder and Liebeskind, 1971) that stimulation at each point may produce analgesia in only a half or a quadrant of the body. The analgesia that is observed is profound: the animals fail to respond to severe noxious stimulation and can even undergo abdominal surgery without signs of pain or discomfort. The analgesia, moreover, may outlast the period of stimulation by as long as five minutes. These animals are not paralysed; rather, they are rendered analgesic without affecting other sensory, motor and cognitive processes.

The relevant points (Figure 31) are distributed in a fairly extensive region of the brainstem as well as portions of the limbic system, but are found primarily within or near the central tegmental tract and adjacent central grey area. Fibres

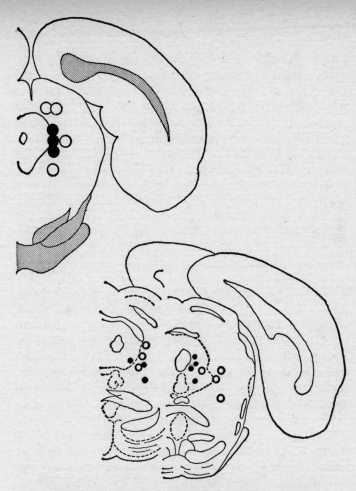

Figure 31 Brainstem sites in the rat which produce profound analgesia when they are electrically stimulated. The filled circles indicate effective sites; the open circles show non-effective sites. *Top:* effective sites reported by Reynolds (1970). Analgesia in this study was defined as the elimination of responsiveness to heavy pressure on the paws and tail. The analgesia was so deep that these animals also underwent abdominal surgery (laparotomy) without showing any signs of pain. *Bottom:* effective sites reported by Mayer, Wolfle, Akil, Carder and Liebeskind (1971, p. 1351). These animals showed no response to intense pinch, heat that produced blistering of the skin, or an ice cold bath that produced escape thirty seconds after the stimulation was turned off.

within this region, it has long been known (Papez and Stotler, 1940; Verhaart, 1949), project to structures higher in the brain as well as to the spinal cord. There is also evidence (Herz, Albus, Metys, Schubert and Teschemacher, 1970) that neurons in this area are part of a larger neural system that is selectively activated by morphine. Indeed, the effects of morphine on transmission of sensory information through the spinal cord are abolished if the cord is cut below the level of the brainstem (Satoh and Takagi, 1971). It is possible, in other words, that morphine exerts at least part of its analgesic effects by exciting fibres that have an inhibitory control over somatic input. The widespread inhibitory influence of this area is further indicated by observations (Jasper and Koyama, 1972) that stimulation of an area in this region produces a heavy release at the cortex of gamma-amino-butyric-acid (GABA), which is believed to be one of the chemical transmitters released by inhibitory neurons. These data, taken together, suggest strongly that an area exists in the brainstem reticular formation that is capable of exerting a powerful inhibitory influence on transmission at all levels of the somatosensory system. It could exert this influence not only on spinal gating mechanisms (either directly on T cells or via the substantia gelatinosa) but at every synaptic level at which information selection and abstraction occurs.

These data provide the basis of the concept (Melzack, 1971, 1972) of a *central biasing mechanism*. The concept proposes that a portion of the brainstem reticular formation acts as a central biasing mechanism (Figure 32) by exerting a tonic inhibitory influence, or bias, on transmission at all synaptic levels of the somatic projection system, including the spinal gating mechanism. The modulation of this influence could therefore play an important role within the framework of the gate-control theory of pain.

Implications of the gate-control theory

The concept of interacting gate-control and action systems can account for the hyperalgesia, spontaneous pain, and many other properties characteristic of pathological pain syndromes

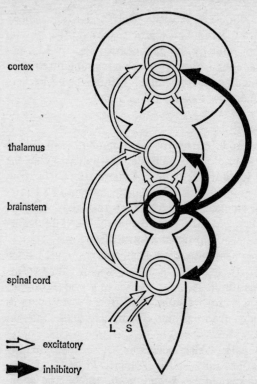

cortex

thalamus

brainstem

spinal cord

L S

➪ excitatory

➡ inhibitory

Figure 32 Schematic diagram of the central biasing mechanism.
Large and small fibres from a limb activate a neuron pool in the
spinal cord, which excites neuron pools at successively higher
levels. The central biasing mechanism, represented by the
inhibitory projection system that originates in the brainstem
reticular formation, modulates activity at all levels. Loss of
inputs to the system would weaken the inhibition; increased
sensory input or direct electrical stimulation would increase the
inhibition. L, large fibres; S, small fibres.
(from Melzack, 1971, p. 409)

(Melzack and Wall, 1965). The occurrence of hyperalgesia
would require two conditions: enough conducting peripheral
axons to generate a T-cell output that can activate the action
system; and a marked loss of the large peripheral nerve fibres,
which may occur after peripheral-nerve lesions or in some of

the neuralgias (Noordenbos, 1959; Kerr and Miller, 1966). Since most of the large fibres are destroyed, the normal inhibition of the input by the gate-control system does not occur. Thus, the input arriving over the remaining myelinated and unmyelinated fibres is transmitted through the unchecked, open gate produced by the small fibre input.

Summation mechanisms

The open gate produced by the predominantly small-fibre input in many pathological pain states would provide the conditions for spatial and temporal summation. The convergence of nerve impulses from the skin, viscera or muscles on to the T-cells would contribute to their total output. In the absence of inhibitory control after the initial T-cell discharge, successive stimuli would produce a more intense and prolonged barrage of nerve impulses after each presentation (Mendell and Wall, 1965). These mechanisms may account for the fact that non-noxious stimuli, such as a series of gentle touches or applications of a warm test-tube, can trigger severe pain in patients suffering phantom limb pain, causalgia and the neuralgias.

Spontaneous pain, which occurs in the absence of any obvious stimulation, can also be explained in terms of summation mechanisms (Melzack and Wall, 1965). The spontaneous activity generated by the remaining small fibres after a nerve lesion would have the effect of keeping the gate open. Low-level, random, ongoing activity would then be transmitted relatively unchecked; the summation of these impulses would produce spontaneous pain. These mechanisms can also account for the long delays between stimulation and pain experience that are frequently observed after peripheral-nerve or dorsal-root lesions. Because the total number of peripheral fibres is reduced, it may take considerable time for the T cells to reach the firing level necessary to trigger pain, so that pain perception and response are delayed.

The role of the sympathetic nervous system in pathological pain can also be understood in terms of the model. Sensory fibres from the viscera are known to travel through the sympathetic ganglia and converge onto the same spinal cord cells that receive cutaneous inputs (Pomeranz, Wall and Weber, 1968; Selzer and Spencer, 1969). Moreover, sympathetic efferent fibres produce changes in blood circulation and sweating, which are a source of tonic input. Sympathectomy, then, would diminish or remove a tonic discharge from viscera, blood vessels, and other deep tissues that could summate with the cutaneous input to produce pain.

The sensory mechanisms alone fail to account for the fact that nerve lesions do not always produce pain and that, when they do, the pain is usually not continuous. Melzack and Wall propose that the presence or absence of pain is determined by the balance between the sensory and the central inputs to the gate-control system. In addition to the sensory influences on the gate-control system, there is a tonic inhibitory influence from the brain. Thus, any lesion that impairs the normal downflow of impulses to the gate-control system would open the gate. Central nervous system lesions associated with hyperalgesia and spontaneous pain (Head, 1920) could have this effect. On the other hand, any condition that increases the flow of descending impulses would tend to close the gate. Since the tonic inhibitory influence exerted by the central biasing mechanism is partly maintained by somatic input, an increase in the T-cell output after a nerve lesion would increase the inhibition exerted at all levels of the somatic projection system. A peripheral nerve lesion, then, would have the *direct* effect of opening the gate, and the *indirect* effect, by increasing T-cell firing and thereby acting on the central biasing mechanism, of closing the gate. The balance between sensory facilitation and central inhibition of the input after peripheral-nerve lesion would account for the variability of pain even in cases of severe lesion.

The gate-control theory also suggests that psychological processes such as past experience, attention, and emotion may influence pain perception and response by acting on the spinal gating mechanism. Some of these psychological activities may open the gate while others may close it. A woman who one day discovers a lump in her breast, and is worried that it may be cancerous, may suddenly feel pain in the breast. If anxiety is prolonged, the pain may increase in severity and even spread to the shoulder and arm. Later, the mere verbal assurance from her doctor that the lump is of no consequence usually produces sudden, total relief of pain.

In some cases, the presence of pain may be determined to a large extent by personal psychological needs. These may range from need for attention from members of the family to masochistic needs for punishment of real or imagined misdeeds. The model suggests that the pain in these cases is determined by psychological processes that act on the gate-control system. They may produce facilitation of all inputs from an area so that nerve impulses generated by pressure or thermal stimuli, or by circulatory activities within the tissue, will be summated to evoke pain (Szasz, 1968). In contrast, they may close the gate to all inputs from a selected body area. Thus, a person may develop 'glove anaesthesia' – total loss of sensation, including pain, from the whole hand. The pattern of sensory loss makes it apparent that there is no neuropathology to account for the symptoms (Walters, 1961). Psychotherapy may relieve the glove anaesthesia, only for it to return months later, perhaps even at the other hand.

Referred pain

The gate-control theory also provides an explanation for many of the phenomena of referred pain. It has long been known that patients with cardiac disease frequently develop pain – which is commonly called 'referred pain' – in the shoulder and upper chest. Examination of cardiac patients by Kennard and Haugen (1955) revealed that most of them show a common pattern of trigger areas (Figure 33) in the shoulder and chest. Pressure on the trigger areas often produces intense pain that may last for hours. Astonishingly, similar examination of a group of subjects who did not have heart disease revealed an almost identical distribution of trigger areas. The application of pressure produced marked discomfort which sometimes lasted for several minutes and even increased in intensity for a few seconds *after* removal of the stimulus.

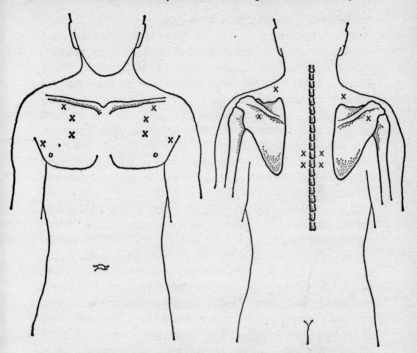

Figure 33 Kennard and Haugen's (1955, p. 297) chart of trigger spots (marked by Xs) in cardiac patients, showing the areas that are most frequently sensitive. Firm pressure on the trigger spots produces discrete, stabbing, 'hot' pain that sometimes persists for as long as several hours. Pressure at the same spots in non-cardiac patients often produces mild pain for several minutes.

The patterns of referred pain are so consistent from person to person that physicians often diagnose the diseased structure on the basis of the pain pattern. It is not surprising, therefore, that within each area of referred pain there are often one or more small trigger zones that are located in more-or-less the same place in most people (Travell and Rinzler, 1946, 1952). Pressure on these trigger zones evokes pain in the referred area and, usually, pain in the related diseased visceral structure. Even more remarkable is the fact that injection of anaesthetic drugs like novocaine in the trigger zones removes the referred pain and, very often, the pain of the diseased viscera. The frequency of painful attacks may decrease significantly after a single such injection. Sometimes, the pain may disappear permanently (Travell and Rinzler, 1946). There is a characteristic sequence to many of these referred pains. For example, in some cases of chronic coronary insufficiency which produces anginal pain, the pain referred to the arm and shoulder becomes the outstanding symptom. The person protects the arm and tends to keep it in a rigid, fixed position. After novocaine injection of the trigger zones, the pain relief allows the patient to use the arm and shoulder normally.

These observations are part of a large body of data on painful trigger areas associated with muscles and the fibrous membrane (fascia) that covers them (Figure 34). These myofascial trigger areas, particularly those found in the lower regions of the back, are sometimes associated with definite nodules of fibrous tissue. Kennard and Haugen (1955) have reviewed evidence that these nodules may develop after virus infections and other fever-producing diseases. Copeman and Ackerman (1947) believe that the trigger zones result from some sort of inflammatory process which produces abnormal fibrous nodules that remain for years. Korr, Thomas and Wright (1955) propose an alternative hypothesis. They suggest that trigger zones develop during the course of growth as a result of musculo-skeletal stresses-and-strains, particularly associated with the muscles of the back. These trigger areas, they observed, are characterized by several kinds of abnormal physiological activity. It is apparent from Kennard and Haugen's study that trigger zones at the chest and back are rare in infants and common in adults, which may be attributable to either inflammation of tissues as a result of disease or to strains on the musculo-skeletal system. The widespread distribution of trigger zones associated with referred pain patterns suggests that both mechanisms may play a role. It is also possible, as Kennard and Haugen have suggested, that the areas at which blood vessels and nerves lie close to the surface, rather than hidden under muscles and other tissues,

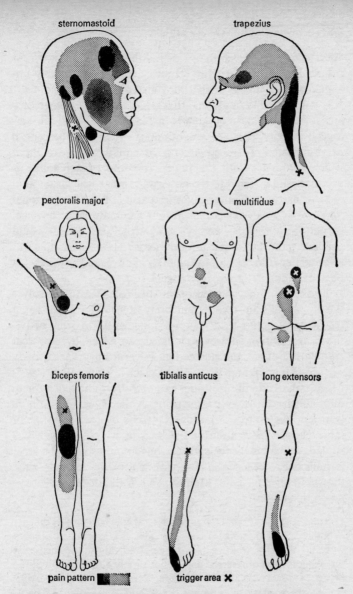

Figure 34 Typical myofascial pain patterns and their related trigger areas reported by Travell and Rinzler (1952, p. 425). When the referred pain pattern of a muscle is known, it can be used to locate the muscle that is the source of pain. The name of the muscle associated with each pain pattern is shown.

are particularly susceptible sites for the formation of trigger zones. The genesis of trigger zones, then, is still largely unexplained, but it is clear that they are commonly found.

It is reasonable to assume that trigger zones, whether they are fibrous nodules or simply areas of abnormal physiological activity, produce a continuous input into the central nervous system. Diseased viscera, then, may produce an input which summates with the input from the trigger zones to produce pain referred to the larger skin areas which surround the trigger zones. Conversely, stimulation of the trigger zones may evoke volleys of impulses that summate with low-level inputs from the diseased visceral structure, which would produce pain that is felt in both areas. These phenomena of referred pain, then, point to summation mechanisms which can be understood in terms of the model.

Two types of mechanisms may play a role. The first involves the spread of pain to adjacent body areas. The T cell has a restricted receptive field which dominates its 'normal activities': in addition, however, it is also affected by electrical stimulation of afferent nerves that cover a much larger body surface (Mendell and Wall, 1965). Melzack and Wall (1965) suggest that this diffuse input is normally inhibited by gate mechanisms, but may trigger firing in the T cell if input is sufficiently intense or if the gate is opened by a selective loss of large fibres. Anaesthesia of the area to which the pain has spread, which blocks the spontaneous impulses from the area, is sufficient to reduce the bombardment of the cell below the threshold level for pain. Melzack and Wall (1970, p. 24) have noted that:

The recent discovery that the small visceral afferents project directly or indirectly onto lamina 5 cells provides the gate-control theory with still further power in explaining referred pain. It is evident that the phenomenon of referred pain is not simply a mislocation of the origin of a visceral afferent barrage. Somewhere in the nervous system there must be a convergence and summation of nerve impulses from the diseased viscera and from the area of skin to which the pain is referred. The pain is exaggerated if skin is touched in the area where the pain is located. Local anaesthesia of skin to which pain is referred abolishes or diminishes the pain. Many theories have suggested possible locations for the convergence between cutaneous and visceral afferents. Lamina 5 cells exhibit this convergence and are monosynaptically connected to visceral afferents. They are therefore good candidates for explaining the phenomenon of referred pain as well as pain of direct cutaneous origin. Both inhibitory and excitatory interactions exist between the converging visceral and cutaneous inputs, which would account for both inhibitory and excitatory interactions at the clinical perceptual level, although the particular conditions necessary for each is not yet clear.

The second mechanism to explain referred pain involves the spread of pain and trigger zones to regions at a considerable distance (Figure 5, p. 56), and may include visceral structures as well as cutaneous and myofascial areas. These referred pains suggest that the gate can be opened by activities in distant body areas. This possibility is consistent with the gate model, since the substantia gelatinosa at any level receives inputs from both sides of the body and (by way of Lissauer's tract) from the substantia gelatinosa in neighbouring body segments. Mechanisms such as these may explain the observations that anginal pain, or pressure on other body areas such as the back of the head may trigger pain in the phantom limb.

Referred pains may also occur after lesions of the central nervous system. Nathan (1956) studied patients who had undergone unilateral or bilateral cordotomy, mostly for the relief of cancer pain, and found that pin pricks applied to analgesic parts of the body such as the leg evoked pain that was felt at distant sites on the same or opposite side of the body (Figure 35). He also observed that, in some patients, the pain was referred to the site of an earlier injury. Nathan's suggestion that these referred pains are due to a hyperexcitability of spinal neurons as a result of the cordotomy is consistent with the concept of a central biasing mechanism. It is possible that the decreased input to the reticular formation after cordotomy brings about a release from inhibition at all levels of the spinal cord. Consequently, the impulses produced by noxious stimulation at analgesic areas would tend, via the substantia gelatinosa, to open the gates at other levels. It is

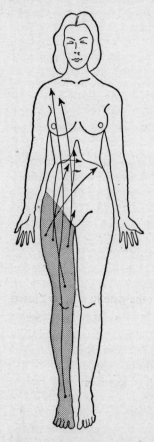

Figure 35 Patterns of referred sensation after cordotomy. The stippled area shows the region of analgesia produced by cordotomy in this woman. Heavy pressure applied to the analgesic skin produced 'an unpleasant form of tingling' that was felt at a non-analgesic part of the body. The sites of stimulation are indicated by dots, and the arrow from each dot indicates the point to which sensation was referred. (from Nathan, 1956, p. 88)

conceivable that the low-level inputs from trigger zones or sites of earlier injury at distant areas of the body would then summate to produce the level of T-cell firing necessary for the perception of pain.

Prolonged pain

The gate-control theory is able to explain some kinds of prolonged pain after peripheral nerve lesion or similar nerve pathology. The loss of large fibres would tend to keep the gate open, and provide the basis for persistent pain. Furthermore, the pain would tend to limit movement of the affected area, which in turn would decrease the normal firing patterns produced by motor activities. A single anaesthetic block could bring about prolonged relief of pain. It would reduce the input so that the total firing level of the T cells would fall below the critical level and stop the pain. As a consequence, the person would move normally and produce normal patterns of input from muscles and other tissues. These patterns, which would include a higher proportion of impulses from the large fibres, would tend to maintain the gate in a more closed position and prevent pain from returning.

This explanation, however, is not always satisfactory. For example, teeth that have been drilled and filled without local anaesthetic may be the site of referred pain when the nasal sinuses are stimulated as long as seventy days later (Reynolds and Hutchins, 1948). A single anaesthetic block of the appropriate nerve from the jaw abolishes the phenomenon. This effect cannot be attributed to a change in the large-to-small fibre ratio or in the descending inhibitory control. Nor can it be due to a chronic local irritation after the dental manipulation: the anaesthetic block could not have affected the teeth themselves. The effect, instead, points to prolonged changes in central neural activity which may be initiated by a brief, painful input and stopped permanently by a single anaesthetic block.

There are many clinical observations which lead to the same conclusion. Momentary pressure on trigger areas in cardiac patients produces severe pain for several hours, and a single anaesthetic block of the areas may abolish recurrent cardiac pain for days, weeks, or longer. How can a single, brief input produce such long effects, and how can a temporary block of input stop it? Similarly, injections of anaesthetic drugs into trigger areas or sympathetic ganglia in people suffering phantom limb pain may produce pain relief that long outlasts the duration of anaesthesia, even though the stump is used little more than it had been prior to treatment.

The explanation of these prolonged effects, which are among the most puzzling features of pain, requires an assumption that is justified by the behavioural and physiological evidence (see chapter 4) on prolonged neural activity in the central nervous system. It is reasonable to assume that prolonged, intense, or otherwise abnormal somatic input produces long-term changes in activity in the central nervous system. It is proposed, therefore, that chronic low-level inputs or brief intense inputs may produce prolonged central neural changes. These neural changes may activate the T cells to produce persistent pain. Conceivably, these self-sustaining neural activities may act like 'trigger points' within the central nervous system itself. They may become a source of tonic activity that rarely exceeds the critical level to evoke pain, but may readily summate with impulses from other sources. Any trauma at a later time may produce an input of sufficiently high level to summate with this continuous, low-level activity and thereby produce persistent, chronic pain.

It is also possible that the open gate produced by nerve lesions or other pathological neural changes may facilitate the occurrence of prolonged central neural activity. That is, the decreased inhibition, whether due to abnormal sensory or central conditions, may increase the probability of self-sustaining activity in neuron pools. Figure 36 is an hypothetical model to account for prolonged activity after release from inhibition. Impulses in axon S normally excite neuron T and the neuronal side chain W-X-Y. Excitation of T produces repetitive discharges in the recurrent inhibitory loop T-Z, followed rapidly by inhibition of T by activity in the side chain. Any disruption in the inhibitory side chain W-X-Y, however,

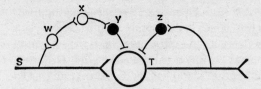

Figure 36 Schematic model to account for prolonged activity. Impulses in axon S normally excite neuron T and the neuronal side chain W–X–Y. Excitation of T produces repetitive discharges in the recurrent inhibitory loop T–Z, followed rapidly by inhibition of T by activity in the side chain. Moderate doses of anaesthetic drugs selectively block the vulnerable multi-synaptic side chain, so that excitation of T leads to prolonged activity in the T–Z loop. Additional anaesthetic doses block the T–Z loop, and the prolonged activity ceases.

would permit the self-sustaining activity to persist indefinitely. The inhibitory influence could, presumably, be exerted by the substantia gelatinosa. A release from inhibition as a result of a selective loss of large fibres (after a peripheral nerve lesion) or a decrease in descending inhibition would provide the conditions for self-sustaining activity. Decreased inhibition from the central biasing mechanism could also produce prolonged activity. In either case, pain would be produced when the total T-cell output exceeds a critical level.

The concept of prolonged, self-sustaining activity can explain the fact that one or more anaesthetic blocks of tender skin areas, trigger points, peripheral nerves, or sympathetic ganglia may produce prolonged relief of phantom limb pain, causalgia and the neuralgias. Anaesthetic block of sensory input for several hours would bring about a cessation of activity in the closed, self-sustaining neuron loops, and would produce pain relief that outlasts the duration of anaesthesia. The relief of pain, moreover, would permit increased motor activity which, in turn, would produce patterned input – particularly in large fibres – that tend to close the gate and prevent or delay the recurrence of sustained activity.

In contrast, the relief of pain after intense stimulation such as local injections of hypertonic saline, or percussion of stump tissues would be due to the increased level of inhibition produced by the input. Since these inputs would excite high-threshold, small-diameter receptor-fibre units, they would open the spinal gate and might evoke severe pain. However, because the T-cell output also projects to the central biasing mechanism, it would raise the level of inhibition and close the gate to subsequent inputs. In addition, the increased inhibition would disrupt the self-sustaining activity at all levels and produce prolonged pain relief.

The concept of a release from inhibition is consistent with the fact that phantom limb pain often resembles the pain felt before amputation. It is possible that prolonged pain may leave 'memory' traces in the somesthetic system (Livingston, 1943), perhaps in the form of closed neuron loops, which are normally inhibited. The release from inhibition as a result of peripheral nerve lesion could bring about activation of the traces to produce persistent, severe pain.

The concept of self-sustaining activity after a release from inhibition can also explain the astonishing observations that an anaesthetic block of the lower spinal cord in amputees sometimes produces severe pain in the phantom leg, even in patients who had previously suffered little or no pain (Moore, 1946; Leatherdale, 1956). If additional anaesthetic is then injected into the sympathetic ganglia or at a higher cord level, the pain may disappear (De Jong and Cullen, 1963). It is possible that the decreased somatic input into the central biasing mechanism, after the first anaesthetic block, would lower the level of inhibition and increase the probability of sustained activity in spinal neuron pools above the level of the block. Because the T cells in these pools may have large receptive fields that normally include fibres from the legs, they would be expected to act as central 'trigger points' for referred pain in the phantom leg. Their increased output would evoke phantom limb pain. The pain would disappear when the T-cell output is lowered below the critical level by abolishing the sympathetic contribution to the T-cell activity or by raising the level of the block to include all the T cells that receive afferent fibres from the legs.

Hyperstimulation analgesia

It is well known that intense somatic stimulation sometimes produces relief of pain for variable periods of time. This type of pain relief, which may be generally labelled as *hyperstimulation analgesia*, is one of the oldest methods used for the control of pain. It is sometimes known as 'counter-irritation', and includes such methods of folk-medicine as application of mustard plasters, ice packs, or hot water bottles to parts of the body. Some of these methods are still frequently used although there has not been (until recently) any theoretical or physiological explanation for their effectiveness. Suggestion and distraction of attention are the usual mechanisms invoked, but neither seems capable of explaining the power of the methods or the long duration of the relief they may afford.

There is, in fact, considerable evidence to show that brief, mildly painful stimulation is capable of bringing about substantial relief of more severe pathological pain for durations that long outlast the period of stimulation. We have already noted (chapter 3) that injection of hypertonic saline into the tissues of the back may produce a sharp brief pain followed by prolonged relief of phantom limb pain. Saline injections into the stump may have the same effect. There is also experimental evidence that one pain may produce a marked rise in threshold to other types of pain. Application of painful cold to the shin of either leg brings about a 30 per cent rise in threshold to pain produced by electrical stimulation of the teeth (Parsons and Goetzl, 1945). The raised threshold may persist for two hours or more. Similarly, paraplegics who suffer pain have a higher threshold to experimentally evoked pain than paraplegics who are pain-free (Hazouri and Mueller, 1950).

It is also well established that referred pains are frequently relieved by intense stimulation applied to trigger areas. Travell and Rinzler (1952) used intense stimulation to abolish myofascial pain. Injection of anaesthetic agents, they observed, is often effective; but so too, astonishingly, is dry needling of the area – simply moving a needle in and out of the area without injecting any substance! Intense cold applied to the area may be equally effective. This effect was presumed at first to be due to local analgesia produced by the cold; but Travell and Rinzler (1952) have come to assume that it is the intense input itself that relieves the pain, since they classify it together with dry needling and heavy pressure as effective methods for the relief of referred pains.

More recently, hypertonic saline has been injected directly into the fluids surrounding the spinal cord in the attempt to relieve severe pain ranging from cancer pain to the neuralgias. The method produces pain briefly but is sometimes followed by prolonged relief of pain (Hitchcock, 1967; Collins, Juras, Houton and Spruell, 1969). Why this method works is not clear, but one possibility is that it produces intense stimulation of spinal cord neurons. Taken together, then, there seems to be a growing body of evidence to show that intense, painful stimulation of brief duration may bring about prolonged, sometimes permanent pain relief.

Neural mechanisms. These observations can be explained by the concept of a *central biasing mechanism* that inhibits transmission through the dorsal horns as well as at higher levels in the somatic projection system. Intense somatic stimuli, of almost any kind, would produce a high level of T-cell output. This may exceed the critical level and produce pain but would also activate the central biasing mechanism which would decrease the T-cell firing level. In some cases, it would block self-exciting neuron loops and produce prolonged relief of pain. It is assumed here that the inhibitory influence exerted by intense somatic stimulation is produced indirectly through the central biasing mechanism. However, this does not preclude direct inhibition at the level of the spinal cord. Wagman and Price (1969) have found that the spontaneous or evoked activity of cells in lamina 5, whose receptive fields cover part or all of one of the legs, can be inhibited by intense stimulation of the opposite leg or even the hands (Figure 37). The short latencies of onset of the effect suggest that it may occur entirely by means of connections in the spinal cord. It is most likely, therefore, that both spinal and supraspinal mechanisms mediate the complex effects of intense stimulation on pain.

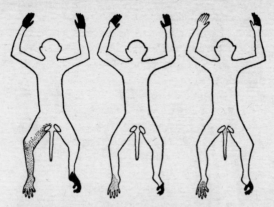

Figure 37 Excitatory and inhibitory receptive fields of dorsal horn cells in the monkey. The excitatory fields of three typical cells are indicated by the stippled areas. The inhibitory fields are shown in black. Large, medium, and small excitatory fields are illustrated from left to right. The inhibition of spontaneous or evoked activity was produced only by *intense* stimulation in the inhibitory fields, and persisted for as long as 1·2 seconds after stimulation was stopped.
(from Wagman and Price, 1969, p. 803)

Acupuncture analgesia. The practice of acupuncture has recently opened the possibility of a unique and remarkable approach to surgical analgesia. Acupuncture refers to the technique, which originated in ancient China, of placing long, fine needles into particular sites at the skin in order to cure various disorders (Figure 38). In recent years, acupuncture has been used to induce a state of profound analgesia so that even major operations on the abdomen, chest or head can be carried out in the totally awake patient. No systematic studies of acupuncture analgesia have yet been reported, so that it is sometimes assumed that the phenomenon is nothing more than hypnosis or strong suggestion. The meagre evidence that is available, however, runs counter to this interpretation. Hypnotic analgesia for major surgical operations can be achieved by highly competent, professional hypnotists in only about 20 per cent of people (LeCron, 1956). In contrast, the available evidence (Brown, 1972) suggests that a much higher

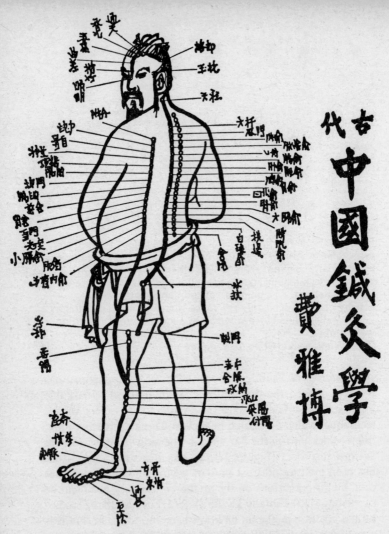

Figure 38 Typical acupuncture charts, showing the sites for insertion of acupuncture needles. Some of the major 'meridians' of the body and their associated internal organs are shown. After two or more acupuncture needles are inserted at selected sites, electrical current is passed through the needles for a period of about twenty minutes. The analgesia that is produced by the acupuncture procedure is reported to be sufficient to permit major surgery.

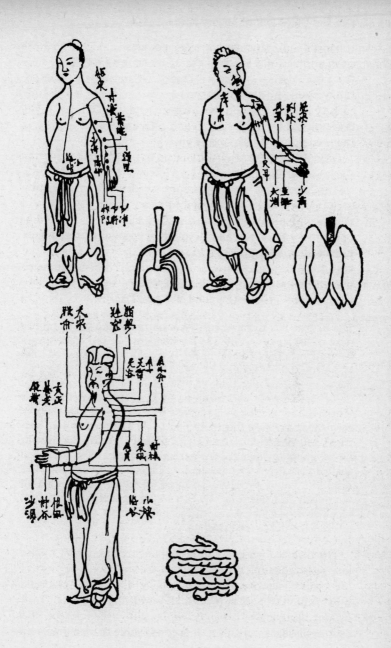

proportion of patients in China, perhaps as high as 90 per cent, undergo surgery with the acupuncture procedure. Furthermore, patients who are deeply hypnotized for surgery rarely speak or act spontaneously. Yet patients undergoing surgery with acupuncture analgesia chat amiably with the doctors, eat pieces of orange, and even show keen interest in the proceedings of the operation.

One of the most fascinating aspects of acupuncture is the site of insertion of needles for different operations. For a thyroid operation, an acupuncture needle was inserted into each forearm, at a point about four inches above the wrist, and at a depth of about one inch (Dimond, 1971). A similar operation was carried out in another hospital with acupuncture needles in the neck and the backs of the wrists (Signer and Galston, 1972). An operation for the removal of the stomach was carried out with four acupuncture needles inserted into the pinna of each ear (Dimond, 1971, p. 1560):

The patient was a slender fifty-year-old man with a non-healing ulcer of the lesser curvature of the stomach. The procedure was to be a gastrectomy (removal of the stomach). This patient had not had medication at bedtime the previous night. He was given sixty mg of meperidine hydrochloride (an analgesic drug) in 500 cc of 5 per cent dextrose during surgery. Acupuncture anaesthesia was introduced by placing four stainless steel needles in the pinna of each ear at carefully identified points. ... The needles were connected to a phasic direct current battery source, delivering six volts at 150 cycles per minute. The patient remained awake, alert, and chatted throughout the procedure. A subtotal gastric resection was done by skilful surgeons, scrubbed, gowned, and disciplined thoroughly in modern or Western surgical practice. This patient required no additional anaesthesia but did note some sensation associated with visceral traction.

The diverse locations of the acupuncture needles to induce analgesia are perplexing in terms of the traditional concept of the central nervous system as having a discrete segmental organization (Figure 4, p. 52). However, the phenomenon of acupuncture analgesia can be explained by the concept of a central biasing mechanism. It is conceivable that acupuncture analgesia is a special case of *hyperstimulation analgesia*. The stimulation of particular nerves or tissues by needles could bring about an increased input to the central biasing mechanism, which would close the gates to inputs from selected body areas. The cells of the midbrain reticular formation are known to have large receptive fields (Rossi and Zanchetti, 1957) and the electrical stimulation of points within the central tegmental tract-central grey area can produce analgesia in a half or quadrant of the body (Mayer, Wolfle, Akil, Carder and Liebeskind, 1971). It is possible, then, that particular body areas may project especially strongly to some reticular areas, and these, in turn, could bring about a complete analgesic block in a large part of the body.

It is clear that acupuncture analgesia involves fairly intense, continuous stimulation of tissues by the acupuncture needles. Sometimes electrical current is passed between two needles. At other times, the needles are continually twirled by hand, which would stimulate the tissues in which they are embedded. The inputs, over a long period of time, could bring about a heightened activity in the central biasing mechanism and, consequently, analgesia of areas of the body. The input produced by stimulation appears to be the critical factor. The injection of novocaine into the acupuncture points, which anaesthetizes them and prevents them from projecting information to the nervous system, also blocks their ability to induce acupuncture analgesia at a distant site (Dimond, 1971). The onset of analgesia, moreover, is usually not instantaneous but appears to develop slowly. Twenty minutes' stimulation was necessary to produce analgesia in one case (Signer and Galston, 1972). Furthermore, post-surgical pain appears not to present a problem because analgesia is reported to outlast the duration of acupuncture stimulation by periods as long as several hours (Brown, 1972).

We still have only fragmentary information about acupuncture analgesia, and there are many unanswered questions. For example, the technique is not used routinely for all patients. Which patients are selected and why are the others rejected? The selection of patients is not surprising: even morphine is effective for only about 75 per cent of cases of post-surgical pain (Beecher, 1959). Placebos, moreover, are effective for 35 per cent of cases – that is, half of the effectiveness of morphine may be considered as a placebo effect (Beecher, 1959). It is important to know, then, whether faith in acupuncture is necessary for the method to work. We have already seen (chapter 2) that a combination of two methods (one of which may be strong suggestion) may produce effective pain relief although each one alone may have no effect. Many of the patients also get small doses of conventional analgesic drugs, which may have tranquillizing or euphoric effects. These too may interact with the effects produced by the acupuncture needles. Fortunately, the increased exchange of scientific information between China and other countries will undoubtedly permit a greater understanding of acupuncture analgesia. Its potential benefits, particularly for surgery in the elderly patient, are very great. It is also certain that the research it will stimulate will provide valuable clues about pain mechanisms generally.

7 The Control of Pain

The gate-control theory has important implications for the treatment of pain (Melzack and Wall, 1965). It provides a new conceptual approach to pain therapy and has suggested new forms of treatment. Furthermore, some of the older methods of therapy (such as successive anaesthetic blocks and the use of 'counter-irritation') take on new meaning in terms of the theory. The gate theory, in recent years, has opened the way for a search for techniques to modulate the sensory input. It suggests that pain control may be achieved by the enhancement of normal physiological activities rather than their disruption by destructive, irreversible lesions. In particular, it has led to attempts to control pain by activation of inhibitory mechanisms.

The need for new approaches to control pain is especially evident in recent assessments of the effectiveness of neurosurgical techniques. Because specificity theory has dominated the field of pain in this century, the natural outcome has been the development of techniques aimed at cutting the so-called pain pathway. When failures occurred, they were attributed to an escape of 'pain fibres', so that operations were carried out at successively higher levels. Generally, the results, particularly for causalgia, phantom limb pain, and the neuralgias, have been disappointing (Sunderland and Kelly, 1948, p. 109):

We should remember that many procedures for the relief of pain have had their crowded hour of general enthusiastic adoption only to fade gradually into oblivion. Results were claimed at first for neurotomy, amputation and posterior rhizotomy, which experience could not confirm. In succession 'periarterial' and postganglionic sympathectomy enjoyed each its vogue of acclaim, only to be replaced by preganglionic sympathectomy, and here we find that, on occasion, it is necessary to do repeated operations and progressively wider excisions. We cannot claim that sympathectomy is of universal value . . . .

Operations have been performed for pain at nearly every possible site in the pathway from the peripheral receptors to the sensory cortex, and at every level the story is the same: some encouraging results but a disheartening tendency for the pain to recur. At whatever level pain is attacked, the impression is conveyed that the whole nervous system makes a coordinated effort to re-establish the pathway. Livingston (1943) was impressed 'with its remarkable ability to find a new route when the customary channels have been blocked'. Leriche (1939) was also discouraged by the failure of surgical attacks to relieve pain: 'nerves are not made to be divided, a demonstration . . . which surgery was long in discovering, and which is not even yet universally admitted as an established fact.'

Repeated section of peripheral nerve fibres means in effect the repeated section of axon cylinders of surviving sensory cells, with the death of some and a spread of disorganization amongst internuncial neurons. Rhizotomy means the degeneration of the central fibres, with transneuronal effects upon the grey matter. Those who do large numbers of antero-lateral cordotomies usually give evidence of diminishing conservatism; increasingly wide sweeps of the knife are advocated in order to be certain of interruption of all the pain fibres. Sometimes bilateral cordotomy is advised.

The use of stereotaxic methods to make lesions in specific midbrain, thalamic and limbic system structures has similarly evoked enthusiasm and hope that an effective pain-relieving operation has at last been found. But the frequent return of pain has led to increasingly larger lesions and, finally, to disenchantment and caution. Spiegel and Wycis, who pioneered some of these techniques, were led to conclude (1966, pp. 13–15) that:

Based upon our experiences in non-tumour material (pain due to herpes zoster, trauma to posterior roots, . . . thalamic vascular lesions), it should be emphasized that rather extensive lesions of the medial thalamus are necessary if one wishes to influence the pain for longer periods. Even lesions of the intralaminar nuclei including the centrum medianum and encroaching upon the lateral part of the dorsomedial nucleus may not be sufficient in the long run despite initial success .... Our own experiences with operative mesencephalic lesions have shown that, after initial relief of the pain, recurrence may occur, in some cases even several years post-operatively, so that long-range observations are necessary in order to obtain insight into the central pain mechanism and to evaluate the merits of procedures attempting to relieve pain .... In view of the necessity of producing rather extensive lesions for long lasting pain relief, these operations should be applied only as a last resort after all measures have failed ....

Our increasing knowledge of pain mechanisms now makes it clear that cutting the peripheral or central nervous system does not simply stop an input from reaching the brain. Surgical section of a peripheral nerve has multiple effects and consequences. It produces sudden, massive, abnormal input volleys which may have long-lasting neural effects; it permanently disrupts normal input patterning; it may result in abnormal inputs from irritating scars and neuromas; and it destroys channels that may be potentially useful to control pain by input modulation methods. Similar consequences occur after cordotomy, which is perhaps the most common operation to relieve pain. Cordotomy does much more than just destroy pain-signalling fibres. It has multiple effects (Melzack and Wall, 1965): it reduces the total number of responding neurons; it changes the temporal and spatial relationships among all ascending systems; and it affects the descending fibre systems that control transmission from peripheral fibres to dorsal horn cells.

The properties of brain activity also defy simple surgical solutions to pain problems. Pain signals project to widespread parts of the brain. If one area is surgically eliminated, there are others that still continue to receive the input. The nervous system, moreover, is able to form new connections and thereby provide new pathways for the sensory input. This plasticity is demonstrated in studies on the effects of lesions of the central grey matter described in chapter 4. The animals failed to respond to noxious heat a month after the lesions were made; nevertheless they began to respond within a few days after onset of testing with noxious stimuli until they became indistinguishable from normal animals. The nervous system appears to have undergone some kind of reorganization so that the input, blocked from ascending through one pathway, is now projected through another.

It is now amply clear that there is a need for other approaches. The gate-control theory has implications for the pharmacological, sensory and psychological control of pain (Melzack and Wall, 1965, 1970).

Pharmacological control of pain

Pharmacological agents may act at a variety of levels in the nervous system. They may act at the receptor level, at the level of the dorsal horn, or at higher levels such as the brainstem. A given drug may possibly act at all three sites.

Analgesics that act at peripheral receptors would presumably decrease the number of nerve impulses they generate as a result of injury. Several compounds have recently been isolated from injured tissues which appear to be intimately related to pain. Bradykinen is one of these compounds. It is produced by the damaged tissues during the course of inflammation and swelling, and is destroyed by enzymes released by the body. Bradykinen produces severe pain when it is injected under the skin or into the deeper tissues of the body. Lim (1968) has recently discovered that aspirin counteracts the action of bradykinen on skin receptors in addition to acting on neural transmission at higher synaptic levels. The discovery of these compounds opens the way to a search for drugs to block the synthesis of bradykinen by body tissues, to metabolize it more rapidly, or to prevent its action on receptors.

Drugs may also affect the transmission of input at the spinal cord level. The gate-control theory suggests that a better understanding of the substantia gelatinosa may lead to new ways of controlling pain. The resistance of the substantia gelatinosa to nerve-cell stains suggests that its chemistry differs from that of other neural tissue. The determination of the nature of this difference may lead to a search for totally new kinds of chemical agents. Drugs that increase the inhibitory effect on transmission by the substantia gelatinosa may therefore be of particular importance in future attempts to control pain. There is already some evidence of the effects of pharmacological agents on gate-control mechanisms. Mendell and Wall (1964) have shown that one of the physio-logical indices of small-fibre activity – the positive dorsal root potential – is completely abolished by light anaesthetic doses of barbiturate. It therefore appears that the facilitatory effect exerted by the small fibres on the transmission of input from peripheral fibres to T cells is blocked by at least some anaes-thetic drugs. It would permit maximal pre- and post-synaptic inhibition and bring about a reduction of the afferent barrage below the critical level necessary for pain. This possibility is supported by the observation (Hillman and Wall, 1969) that barbiturate anaesthesia diminishes the activity evoked in lamina-5 cells by electrical stimulation of the skin.

The effects of anaesthetic drugs on transmission in the reticular formation is well documented. Nitrous oxide, at analgesic levels, strikingly diminishes the amplitude of potentials evoked in the midbrain reticular formation by supramaximal stimulation of the tooth pulp (Haugen and Melzack, 1957). The powerful effects of barbiturates and other anaesthetics at this level have been described and evaluated by French, Verzeano and Magoun (1953). Part at least of the effects on pain may be the prevention of summation of sensory inputs at the brainstem level so that the critical level necessary to trigger pain is not exceeded.

Besides analgesic and anaesthetic drugs that block trans-mission in sensory projection pathways, there may be pharma-cological agents that have selective effects on other aspects of pain processes. Tegretol, as we have seen (chapter 3), stops pain in many cases of tic douloureux. It is also effective for some kinds of burning, paroxysmal pains that result from severe spinal-cord lesions (Gibson and White, 1971). Tegretol may not exert a direct blocking action on transmission path-ways but may act selectively on neural networks that underlie prolonged abnormal activity in the central nervous system. Too little is known about the action of the drug to warrant further speculation. But it suggests the existence of classes of drugs that may so far have no suspected analgesic properties yet may provide powerful new pharmacological approaches to pain control.

Sensory control of pain

The gate theory suggests three general methods to control pain by modulation of the input:

1. The use of anaesthetic blocking agents to decrease the number of nerve impulses that impinge on the T cells.

2. Low-level stimulation which selectively activates the large fibres that inhibit transmission from peripheral fibres to T cells.

3. Intense stimulation which activates brainstem mechanisms that exert an inhibitory influence on the spinal gate-control system and at higher synaptic transmission levels.

Anaesthetic blocks of the sensory input often produce pain relief that outlasts the duration of the blocks (Livingston, 1943; Kibler and Nathan, 1960). Successive blocks may relieve pain for increasingly long periods of time. Anaesthetic blocks of tender skin areas, trigger points, peripheral nerves, or sympathetic ganglia would have the effect of diminishing the total sensory input that bombards the T-cells. They would, therefore, reduce the T-cell output below the critical level necessary to evoke pain. These blocks, moreover, could bring about a cessation of activity in closed, self-sustaining neuron loops, so that temporary blocks would produce long periods of relief. Furthermore, the relief of pain would permit increased use of the affected parts of the body, allowing the patient to carry out normal motor activities. These, in turn, would produce patterned inputs (particularly from muscles) that would contain a high proportion of active large fibres that would further close the gate and delay the recurrence of pain.

According to the gate theory, pain relief may also be produced by forms of stimulation that selectively activate the large fibres. Attempts to produce this kind of selective activation by vibration or other tactile stimuli applied to the skin encounter serious difficulties. Mild tactile stimulation tends to excite small as well as large fibres, so that it may be difficult to produce a net inhibitory effect. Although these methods are seemingly simple, their effectiveness appears to depend on complex interactions between the inputs produced by the source of pain and those produced by the stimuli to control it. The control of itch by scratching or vibration provides evidence of these interactions. Vibration, like scratching, decreases the perceived intensity of mild or moderate itch, but may turn severe itch into frank pain (Melzack and Schecter, 1965). Melzack, Wall and Weisz (1963) similarly found that vibration diminishes the pain produced by low levels of shock but enhances the pain evoked by high shock levels. A different effect, however, was observed by Higgins, Tursky and Schwartz (1971) who used an ingenious cuff device to vibrate a larger area of skin. They found that the vibration diminished the pain produced by intense shock but had no effect on pain evoked by low shock levels. In clinical cases, this interaction may also be determined by the nature of the injury. Some kinds of neuralgic pain are evoked by gentle touch but are unaffected by heavy pressure (Kugelberg and Lindblom, 1959). The interactions may be even more complex: vibration may diminish phantom limb pain and some neuralgic pains but make other neuralgic pains worse (Melzack and Bromage, 1972). Many additional factors (such as earlier operations on the nervous system or the duration of pain) may also play roles which have yet to be determined.

A much more effective method to selectively activate large fibres appears to be the direct electrical stimulation of sensory nerves. Wall and Sweet (1967) discovered that stimulation of peripheral nerves with low-intensity shock may abolish neuralgic pain in the arm or hand. The electrical stimulation, which selectively activates the large fibres that are assumed to close the gate, clearly brings about changes for prolonged periods of time. Stimulation for two minutes may abolish severe neuralgic pain for more than two hours (White and Sweet, 1969). Successive periods of stimulation, in one case, produced pain relief lasting for months. The use of analgesic drugs in combination with stimulation permits a further degree of control over pain (White and Sweet, 1969). The fact that even temporary pain relief allows active movement of the hand, with ensuing normal proprioceptive input, may be a decisive factor in producing enduring relief in these cases. The effectiveness of this technique also varies according to the kind of pain. Patients with cancer pain report that the pain is diminished during the period of stimulation but returns shortly after it is stopped. In these cases, continuous stimulation may be necessary to produce significant relief. The value of the method for the relief of causalgic pain is now firmly established. Meyer and Fields (1972) have reported significant relief of pain in six out of eight patients who received electrical stimulation of the affected nerve by means of electrodes placed on the skin above it. Stimulation for two or three minutes produced pain relief for periods ranging from five minutes to ten hours. Equally impressive is the fact that two patients who had not been helped by sympathectomy were relieved of pain by the nerve stimulation. The technique is new and, as White and Sweet (1969) note, it holds great promise as a means for the control of at least some kinds of pain.

Yet another method of electrical stimulation for pain relief has recently been developed. Shealy, Mortimer, and Hagfors (1970) and Nashold and Friedman (1972) have placed electrodes on the surface of the dorsal columns in patients suffering pain as a result of various injuries or diseases. Receivers attached to the electrodes are inserted under the skin, and the patient is given a small transmitter control box in order to stimulate his own dorsal columns to relieve the pain. Effective relief is reported by about 60 per cent of the patients when the stimulation produces a tingling feeling (Nashold and Friedman, 1972). When the procedure is ineffective for a patient, the electrodes can be removed or placed at another site. The data obtained so far suggest that the method may be at least as effective as cordotomy, and has the decided advantage that it does not involve destruction of neural tissue. The theoretical basis for the method is the possibility that the nerve impulses evoked in the dorsal columns by electrical stimulation descend antidromically to the dorsal horns and then orthodromically along branches to the dorsal horn cells. Because the largest somatic fibres are activated, the impulses tend to close the gate (Hillman and Wall, 1969). An alternative explanation is that the dorsal column stimulation produces impulses that eventually project to the central biasing mechanism, thereby increasing inhibition of activity throughout the somatosensory projection system (Melzack, 1971). Whatever the explanation may be, it appears that we may have an exciting new technique for the control of pain. However, the technique has not been used long enough to be certain of its long-range effectiveness.

The concept of a central biasing mechanism suggests that intense sensory stimulation may be able to diminish or abolish pain. The concept is consistent with observations that pain relief may be produced by application of cold or heat, intense pressure, dry needling, or injection of hypertonic saline into tissues. Methods to produce pain relief by 'counter-irritation' have long been part of folk medicine. Because they could not be explained by any of the traditional theories, they were usually given little attention. One of the few systematic studies of these methods (Gammon and Starr, 1941) shows that brief, successive applications of cold or heat produce significant decreases in some kinds of pain. The way these stimuli are applied is an important determinant of their effectiveness.

It is conceivable that the control of pain by sensory modulation may one day be sufficiently well understood to obviate the need for any surgical procedures. Theoretically, it should be possible to mimic the effects of electrical stimulation of nerves or the dorsal columns by stimulating the skin with appropriate patterns of tactile, thermal, or electrical stimulation. Application of these stimuli directly to the skin would eliminate the necessity to expose peripheral nerves or the spinal cord.

Psychological control of pain

The gate-control theory proposes that cognitive activities such as anxiety, attention, and suggestion can influence pain by acting at the earliest levels of sensory transmission. The degree of central control, however, would be determined, in part at least, by the temporal-spatial properties of the input patterns. Some of the most unbearable pains, such as cardiac pain, rise so rapidly in intensity that the patient is unable to achieve any control over them. On the other hand, more slowly rising temporal patterns are susceptible to central control and may allow the patient to 'think about something else' or use other stratagems to keep the pain under control.

It is clear that the search for new approaches to pain therapy might well profit by directing thinking towards the contributions of motivational and cognitive processes. Pain can be treated not only by trying to manipulate the sensory input, but also by influencing motivational and cognitive factors as well. Tranquillizers, muscle-relaxants, suggestion, placebos, and hypnosis are known (Beecher, 1959) to exert a powerful influence on pain, but the historical emphasis on sensory mechanisms and the relative neglect of the motivational and cognitive contributions to pain has made these forms of therapy suspect, seemingly fraudulent, almost a sideshow in the mainstream of pain treatment. Yet, if we can recover from historical accident, these methods deserve more attention than they have received.

Pain mechanisms are complex and it is wrong to expect simple push-button control of pain. But the complexity of pain should not deter us from searching realistically for ways to achieve cognitive control over it. Psychological procedures require effort and time on the part of the patient and the clinician. However, they provide an important approach to pain therapy, particularly for those pain states which cannot be brought under satisfactory control by local anaesthesics or any other techniques. Therapists have persistently sought methods to abolish pain in the way that telephone transmission can be abolished by cutting a wire. But no techniques have such clear-cut results. Perhaps, then, we should not aim at totally abolishing pain but, rather, at reducing it to bearable levels.

There is evidence (Sternbach, 1970) that many psychological

approaches are able to produce some measure of pain relief. They include:

1. Desensitization techniques in which exposure to pain and the opportunity to gain some kind of control over it help relieve it or make it bearable.
2. Hypnotic suggestion techniques (Barber, 1969; Hilgard, 1971).
3. Progressive relaxation methods.
4. Teaching patients to utilize electro-encephalographic or other indices of physiological activity to develop a state of mind which allows them to cope with pain (Gannon and Sternbach, 1971).
5. The use of stratagems to distract attention or give meaning to the situation so that the pain becomes more bearable (Melzack, Weisz and Sprague, 1963).
6. Psychotherapeutic or pharmacological techniques to relieve depression.

Sternbach (1970) indicates that the relief of depression is particularly effective in bringing about increased pain tolerance, but notes that all of these approaches have value in the treatment of pain. These psychological methods may not abolish pain entirely but may decrease some kinds of pain from unbearable to bearable levels, an achievement that has not received the recognition it merits.

Pain clinics and research

Research on the puzzle of pain, as we have seen, goes on at many levels. Clinical observation and research, behavioural and physiological experiments, and the search for new pharmacological compounds all focus on different aspects of a common problem. We now appear to be moving towards a more coordinated effort in the attempt to solve the puzzle and thereby relieve one major kind of human misery.

Presently, few hospitals are organized to cope with the more complex kinds of pain. People suffering severe pain may be transferred from one doctor to another (from neurologist to neurosurgeon and finally psychiatrist) with little or no help. They may cycle through these specialists several times without experiencing any significant pain relief. What is needed is a concerted effort in which new modes of therapy can be attempted and evaluated. In short, what is needed are *pain clinics* in which specialists can work together to deal specifically with pain problems. In such clinics, an interchange of ideas can occur and the conditions are conducive to novel, imaginative approaches. Pain, in such a clinic, is not merely a symptom which each specialist perceives from his own point of view. Rather, it is the pain syndrome that is itself examined, and the integration of many specialties to treat it is more easily achieved. A further merit is that the clinic allows the accumulation of data – such as the relative effectiveness of different therapeutic procedures – that are often lost as the patient visits each specialist in his own clinic. Only a few pain clinics exist. Many more are urgently needed.

There are few problems that are more challenging than the puzzle of pain. Its solution is compelled by the human desire to relieve pain and suffering, both for those who will recover and go on to lead useful lives, and for those whose lives are coming to an end. Many of us do not fear death but rather fear the pain that may precede it. Patients in the last stages of cancer have often come to terms with the knowledge that the end is near. Their worry is that they may not have the courage to bear the pain of their final weeks with the dignity they fought so hard to achieve in daily life. There is no merit to this suffering, no lesson to be learned.

The pain clinic would allow the development of a battery of techniques to control pain. The pharmacological, sensory, and psychological methods of pain control do not exclude each other. A combination of several methods – such as electrical stimulation of nerves and appropriate drugs – may be necessary to provide satisfactory relief. The effective combination may differ for each type of pain, and possibly for each individual, depending on such factors as the patient's earlier medical history, pattern of spread of trigger zones, and the duration of the pain. But it is only in a clinic, where many cases are seen and complete data files are kept, that sufficient experience and knowledge can be acquired to allow the best judgement in each case.

A pain clinic, moreover, would allow the judicious use of drugs for terminal patients. At present, this decision depends largely on the individual physician. One physician may seek any means, even major surgical operations, to avoid administering morphine, presumably out of fear of turning the terminal patient into an addict. Another may decide that a person's final weeks should be spent in tranquillity, and provide drugs such as morphine whenever they are requested by the patient. These are complex social issues and they may be handled best by a group of physicians and scientists who have gained familiarity with the ravages of prolonged severe pain on the human mind.

Pain clinics are obviously not enough. We also need to pursue basic research on the functions of the nervous system, using anatomical, physiological, psychological, and pharmacological approaches. Advances in factual and theoretical knowledge have led to new imaginative approaches to the solution of pain problems. It is reasonable to assume that they will continue to do so.

The puzzle of pain, as we have seen, is far from solved. Increased research on pain is clearly needed. It is astonishing that there are no major research institutes devoted specifically to the study of pain. Furthermore, the number of scientists who work on the problem is small in comparison with the magnitude of its importance. We have a remarkable capacity to forget pains that we have suffered in the past, and it is often difficult to comprehend the suffering of another person. Research time and money is devoted to many problems of obvious clinical significance, but pain, often considered the symptom and not the disease, receives far less attention. It is encouraging to see a renewed interest in pain in recent years, partly because a field that lay conceptually stagnant for almost a century has suddenly become alive – full of new controversy and renewed fascination. The simple-minded answers of the past are no longer accepted so passively, and new questions are constantly being raised that are sure to challenge the young investigator, and lead to new approaches and new understanding. There are few problems more worthy of human endeavour than the relief of pain and suffering.

Glossary

ablation The removal by surgery of any part of the body. In neurosurgery, refers to removal of part of the brain.

afferent fibre Nerve fibre which conducts nerve impulses from a sense organ to the central nervous system, or from lower to higher levels in sensory projection systems in the spinal cord and brain.

anaesthesia Total loss of sensation in all or part of the body.

anaesthetic As an adjective, refers to an area that has lost all sensitivity. As a noun, refers to drugs that induce the total loss of sensitivity either in a localized area or in the whole body after loss of consciousness.

analgesia Loss of sensitivity to pain without loss of other sensory qualities or of consciousness.

analgesic As an adjective, refers to an area that is insensitive to pain. As a noun, refers to any pain-relieving drug.

antidromic Propagation of a nerve impulse along an axon in a direction that is the reverse of the normal direction of transmission.

asymbolia Loss of the ability to appreciate some aspect of the sensory world. *Pain asymbolia:* inability to appreciate pain – that is, feel it in the normal way or grasp its implications.

axon The part of a nerve cell (neuron) which is the essential conducting portion. Often called simply the 'nerve fibre'. Nerve impulses are conducted along the axon in a direction *away* from the cell body.

brainstem The part of the brain that lies between the spinal cord and the cerebral cortex. Generally refers to those parts of the brain called the medulla oblongata, pons, and midbrain. Sometimes it is used to include the thalamus.

central nervous system In mammals, refers to the spinal cord and brain.

clonic From the word 'clonus' referring to rapid alternate contraction and relaxation of a muscle.

commissural fibres A tract of neurons that connects two areas on opposite sides of the brain or spinal cord.

contralateral On the opposite side.

conversion hysteria Transformation of an emotional disturbance into a physical manifestation such as paralysis, anaesthesia of part of the body, or pain.

cortex The outer layer of an organ. Thus, *cerebral cortex:* the layers of nerve cells at the outer part of the brain.

cutaneous Relating to the skin.

decompression The relief of pressure within an organ by means of an operation to release excessive fluid. Thus, *subtemporal decompression:* the release of cerebrospinal fluid or blood through a burr-hole near the temporal (or lower side) part of the skull.

dendrite The part of a nerve cell (neuron) which conducts nerve impulses toward the cell body. Peripheral *sensory* nerve fibres may be considered as exceptionally long, highly specialized forms of dendrites.

dermatome The area of skin innervated by a single sensory root of the spinal cord.

ecchymosis Bruise; bleeding under the skin, usually after injury.

efferent fibre Neuron which conducts nerve impulses away from the central nervous system (to muscles or glands), or from higher to lower areas in the nervous system (such as a neuron that transmits from the brain to the spinal cord).

encephalon The brain. Thus, *encephalopathy:* any disease of the brain.

ephapse An artificial synapse (junction) between two conducting fibres that may occur after injury.

evisceration Removal of viscera (abdominal and thoracic organs).

ganglion An aggregate of nerve cell bodies. Thus, *sympathetic ganglion:* nerve cell bodies associated with the sympathetic nervous system.

hyperaesthetic Excessively sensitive, so that even non-noxious stimuli (such as a light touch) evoke pain.

introspection The analysis, by a person, of the sensory, emotional and other qualities of conscious experience.

ipsilateral On the same side.

jactitations Jerking, paroxysmal movements.

lumbar The part of the back and sides of the body between the lowest pair of ribs and the top of the pelvis.

median nerve One of the three major nerves that supply the hand. The other two are the radial and ulnar nerves. The sensory area innervated by the median nerve is complex but may be described roughly as the middle portion of the hand, particularly the middle and index fingers and the adjacent portions of the thumb and ring fingers.

myelin A fatty substance surrounding nerve fibres, thereby forming an insulating sheath. *Myelinated:* covered by a myelin sheath.

neuroma A nodule of regenerated tissue.

neuron The structural unit of the nervous system, consisting of a nerve cell and its conducting dendrites and axon.

orthodromic Propagation of a nerve impulse in the normal direction; in axons, away from the cell body.

peripheral nerves Bundles of nerve fibres that connect sensory or motor organs to the central nervous system.

pinna The external part of the ear.

placebo Greek word that means 'I please'. Usually a pill or injectable solution of sugar or salt given in place of an analgesic agent.

polysurgical addiction Refers to patients who appear to have a compelling need for surgical operations.

post-tetanic potentiation From 'tetanus', which refers to the continued contraction of a muscle, which can be produced by a rapid succession of electrically excited nerve impulses. Post-tetanic potentiation refers to the enhancement (potentiation) of muscle contractions or of nerve signals in motor neurons after prolonged, intense stimulation of the related sensory root.

proprioceptive Sensory signals from muscles, tendons and joints.

psychophysics Study of the relationship between stimulus intensity and the intensity of the resultant sensory experience.

roentgenography The use of x-rays to reveal internal structure. The name is derived from Wilhelm Roentgen, the discoverer of x-rays.

sacrum The continuation of the backbone below the lumbar vertebrae, consisting of several vertebrae joined together and making up the central bone of the pelvis. Thus, *sacral:* relating to the sacrum.

soma Greek word for 'body'. Somatic (or 'somatosensory') input refers to sensory signals from all tissues of the body, including skin, viscera, muscles or joints.

somaesthesis Sensory experience derived from the body.

subtemporal decompression See *decompression.*

sympathetic nervous system One part of the autonomic nervous system, consisting of a chain of ganglia lying outside and parallel to the spinal cord, and nerve fibres that conduct to viscera, blood vessels and glands.

synapse The relay junction between two neurons. The axon terminals of a neuron release a chemical transmitter that flows across the synapse and influences the dendrites or cell body of an adjacent neuron. The transmitter may excite the cell (or facilitate its excitation by other neurons) or it may inhibit the cell and prevent it from firing (or decrease its firing rate).

thalamus One of the major relay stations of the central nervous system, lying at the top of the brainstem and between the cerebral hemispheres. It relays information projected by the sensory systems to the cortex and by the cortex to motor systems or to other brain areas.

trigeminal nerve The fifth nerve of the head. It carries sensory signals from the skin of the face, parts of the eyes, and a large part of the inner structures and membranes of the mouth and nose.

viscera The specialized internal organs of the abdomen and chest. Singular: *viscus.*

DeJong, R. H., and Cullen, S. C. (1963), 'Theoretical aspects of pain: bizarre pain phenomena during low spinal anesthesia', *Anesthesiology*, vol. 24, p. 628.

Delgado, J. M. R. (1955), 'Cerebral structures involved in the transmission and elaboration of noxious stimulation', *J. Neurophysiol.*, vol. 18, p. 261.

Delgado, J. M. R., Rosvold, H. E., and Looney, E. (1956), 'Evoking conditioned fear by electrical stimulation of subcortical structures in the monkey brain', *J. comp. physiol. Psychol.*, vol. 49, p. 373.

Descartes, R. (1644), *L'homme*, translated by M. Foster, *Lectures on the History of Physiology during the 16th, 17th and 18th Centuries*, Cambridge University Press.

Dick-Read, G. (1962), *Childbirth Without Fear*, Dell.

Dimond, E. G. (1971), 'Acupuncture anaesthesia', *J.A.M.A.*, vol. 218, p. 1558.

Donaldson, H. H. (1885), 'On the temperature sense', *Mind*, vol. 10, p. 399.

Douglas, W. W., and Ritchie, J. M. (1957), 'Non-medullated fibres in the saphenous nerve which signal touch', *J. Physiol.*, vol. 139, p. 385.

Drake, C. G., and McKenzie, K. G. (1953), 'Mesencephalic tractotomy for pain', *J. Neurosurg.*, vol. 10, p. 457.

Echlin, F., Owens, F. M., and Wells, W. L. (1949), 'Observations on "major" and "minor" causalgia', *Arch. Neurol. Psychiat.*, vol. 62, p. 183.

Ewalt, J. R., Randall, G. C., and Morris, H. (1947), 'The phantom limb', *Psychosom. Med.*, vol. 9, p. 118.

Feinstein, B., Luce, J. C., and Langton, J. N. K. (1954), 'The influence of phantom limbs', in P. Klopsteg and P. Wilson (eds.), *Human Limbs and Their Substitutes*, McGraw-Hill.

Foltz, E. L., and White, L. E. (1962), 'Pain "relief" by frontal cingulumotomy', *J. Neurosurg.*, vol. 19, p. 89.

Frankstein, S. A. (1947), 'One unconsidered form of the part played by the nervous system in the development of disease', *Science*, vol. 106, p. 242.

Freeman, W., and Watts, J. W. (1950), *Psychosurgery in the Treatment of Mental Disorders and Intractable Pain*, C. C. Thomas.

French, J. D., Verzeano, M., and Magoun, W. H. (1953), 'Neural basis of anesthetic state', *Arch. Neurol. Psychiat.*, vol. 69, p. 519.

Frey, M. von (1895), 'Beitrage zur Sinnesphysiologie der Haut', *Ber. d. kgl. sächs. Ges. d. Wiss.*, math.-phys. Kl., vol. 47, p. 181.

Gammon, G. D., and Starr, I. (1941), 'Studies on the relief of pain by counter-irritation', *J. clin. Invest.*, vol. 20, p. 13.

GANNON, L., and STERNBACH, R. A. (1971), 'Alpha enhancement as a treatment for pain: a case study', *J. behav. Ther. exper. Psychiat.*, vol. 2, p. 209.

GARDNER, E. D. (1940), 'Decrease in human neurons with age', *Anatomical Record*, vol. 77, p. 529.

GARDNER, W. J., and LICKLIDER, J. C. R. (1959), 'Auditory analgesia in dental operations', *J. Amer. Dent. Assn.*, vol. 59, p. 1144.

GERARD, R. W. (1951), 'The physiology of pain: abnormal neuron states in causalgia and related phenomena', *Anesthesiology*, vol. 12, p. 1.

GIBSON, J. C., and WHITE, L. E. (1971), 'Denervation hyperpathia: a convulsive syndrome of the spinal cord responsive to carabamazepine therapy', *J. Neurosurg.*, vol. 35, p. 287.

GOLDSCHEIDER, A. (1886), 'Zur Dualität des Temperatursinns', *Pflügers Arch. ges. Physiol.*, vol. 39, p. 96.

GOLDSCHEIDER, A. (1894), *Ueber den Schmerz in Physiologischer und Klinischer Hinsicht*, Hirschwald.

GRASTYAN, E., CZOPF, J., ANGYAN, L., and SZABO, I. (1965), 'The significance of subcortical motivational mechanisms in the organization of conditional connections', *Acta Physiol. Acad. Sci. Hung.*, vol. 26, p. 9.

HAGBARTH, K. E., and KERR, D. I. B. (1954), 'Central influences on spinal afferent conduction', *J. Neurophysiol.*, vol. 17, p. 295.

HALL, K. R. L., and STRIDE, E. (1954), 'The varying response to pain in psychiatric disorders: a study in abnormal psychology', *Brit. J. med. Psychol.*, vol. 27, p. 48.

HALLIDAY, A. M., and MINGAY, R. (1961), 'Retroactive raising of a sensory threshold by a contralateral stimulus', *Quart. J. exper. Psychol.*, vol. 13, p. 1.

HARDY, J. D., WOLFF, H. G., and GOODELL, H. (1952), *Pain Sensations and Reactions*, Williams & Wilkins.

HAUGEN, F. P., and MELZACK, R. (1957), 'The effects of nitrous oxide on responses evoked in the brainstem by tooth stimulation', *Anesthesiology*, vol. 18, p. 183.

HAZOURI, L. A., and MUELLER, A. D. (1950), 'Pain threshold studies on paraplegic patients', *Arch. Neurol. Psychiat.*, vol. 64, p. 607.

HEAD, H. (1920), *Studies in Neurology*, Kegan Paul.

HEBB, D. O. (1949), *The Organization of Behaviour*, Wiley.

HEBB, D. O. (1972), *Textbook of Psychology*, Saunders.

HENDERSON, W. R., and SMYTH, G. E. (1948), 'Phantom limbs', *J. Neurol. Neurosurg. Psychiat.*, vol. 11, p. 88.

HERZ, A., ALBUS, K., METYS, J., SCHUBERT, P., and
TESCHEMACHER, H. (1970), 'On the central sites for the
anti-nociceptive action of morphine and fentanyl', *Neuropharmacol.*,
vol. 9, p. 539.

HIGGINS, J. D., TURSKY, B., and SCHWARTZ, G. E. (1971),
'Shock-elicited pain and its reduction by concurrent tactile
stimulation', *Science*, vol. 172, p. 866.

HILGARD, E. R. (1965), *Hypnotic Susceptibility*, Harcourt Brace
& World.

HILGARD, E. R. (1971), 'Hypnotic phenomena: the struggle for
scientific acceptance', *Amer. Scientist*, vol. 59, p. 567.

HILL, H. E., KORNETSKY, C. H., FLANARY, H. G., and
WIKLER, A. (1952a), 'Effects of anxiety and morphine on
discrimination of intensities of painful stimuli', *J. clin. Invest.*,
vol. 31, p. 473.

HILL, H. E., KORNETSKY, C. H., FLANARY, H. G., and
WIKLER, A. (1952b), 'Studies of anxiety associated with
anticipation of pain. I. Effects of morphine', *Arch. Neurol.
Psychiat.*, vol. 67, p. 612.

HILLMAN, P., and WALL, P. D. (1969), 'Inhibitory and excitatory
factors influencing the receptive fields of lamina 5 spinal cord
cells,' *Exper. Brain Res.*, vol. 9, p. 284.

HITCHCOCK, E. (1967), 'Hypothermic subarachnoid irrigation for
intractable pain', *Lancet*, vol. 1, p. 1133.

HONGO, T., JANKOWSKA, E., and LUNDBERG, A. (1968),
'Post-synaptic excitation and inhibition from primary afferents in
neurons in the spinocervical tract', *J. Physiol.*, vol. 199, p. 569.

HUNT, C. C., and MCINTYRE, A. K. (1960), 'Properties of
cutaneous touch receptors in cat', *J. Physiol.*, vol. 153, p. 88.

HUTCHINS, H. C., and REYNOLDS, O. E. (1947), 'Experimental
investigation of the referred pain of aerodontalgia', *J. dent. Res.*,
vol. 26, p. 3.

HUXLEY, A. (1952), *The Devils of Loudon*, Harper.

JASPER, H. H., and KOYAMA, I. (1972), personal communication.

JEWESBURY, E. C. O. (1951), 'Insensitivity to pain', *Brain*, vol. 74,
p. 336.

KALLIO, K. E. (1950), 'Permanency of the results obtained by
sympathetic surgery in the treatment of phantom pain',
Acta Orthop. Scand., vol. 19, p. 391.

KEELE, K. D. (1957), *Anatomies of Pain*, Oxford University Press.

KENNARD, M. A., and HAUGEN, F. P. (1955), 'The relation of
subcutaneous focal sensitivity to referred pain of cardiac origin',
Anesthesiology, vol. 16, p. 297.

KERR, F. W. L., and MILLER, R. H. (1966), 'The ultrastructural
pathology of trigeminal neuralgia', *Arch. Neurol.*, vol. 15, p. 308.

KEYNES, G. (1952), *The Apologie and Treatise of Ambroise Paré*, Chicago University Press.

KIBLER, R. F., and NATHAN, P. W. (1960), 'Relief of pain and paraesthesiae by nerve block distal to a lesion', *J. Neurol. Neurosurg. Psychiat.*, vol. 23, p. 91.

KING, H. E., CLAUSEN, J., and SCARFF, J. E. (1950), 'Cutaneous thresholds for pain before and after unilateral prefrontal lobotomy', *J. nerv. ment. Dis.*, vol. 112, p. 93.

KOLB, L. C. (1954), *The Painful Phantom: Psychology, Physiology and Treatment*, C. C. Thomas.

KORR, I. M., THOMAS, P. E., and WRIGHT, H. M. (1955), 'Symposium on the functional implications of segmental facilitation', *J. Amer. Osteopath. Assn*, vol. 54, p. 1.

KOSAMBI, D. D. (1967), 'Living prehistory in India', *Sci. Amer.*, vol. 216 (February), p. 105.

KROEBER, A. L. (1948), *Anthropology*, Harcourt.

KUGELBERG, E., and LINDBLOM, U. (1959), 'The mechanism of pain in trigeminal neuralgia', *J. Neurol. Neurosurg. Psychiat.*, vol. 22, p. 36.

LAMBERT, W. E., LIBMAN, E., and POSER, E. G. (1960), 'The effect of increased salience of a membership group on pain tolerance', *J. Personality*, vol. 28, p. 350.

LARSELL, O. (1951), *Anatomy of the Nervous System*, Appleton-Century-Crofts.

LASHLEY, K. S. (1951), 'The problem of serial order in behavior', in L. A. Jeffress (ed.), *Cerebral Mechanisms in Behavior*, Wiley.

LEATHERDALE, R. A. L. (1956), 'Phantom limb pain associated with spinal analgesia', *Anaesthesia*, vol. 11, p. 249.

LECRON, L. M. (ed.) (1956), *Experimental Hypnosis*, Macmillan.

LERICHE, R. (1939), *The Surgery of Pain*, Williams & Wilkins.

LESSAC, M. (1965), 'The effects of early isolation and restriction on the later behavior of beagle puppies', Ph.D. thesis, University of Pennsylvania.

LI, C. H., and ELVIDGE, A. R. (1951), 'Observations on phantom limbs in a paraplegic patient', *J. Neurosurg.*, vol. 8, p. 524.

LIM, R. K. S. (1968), 'Neuropharmacology of pain', in R. K. S. Lim (ed.), *Pharmacology of Pain*, Pergamon.

LIVINGSTON, W. K. (1943), *Pain Mechanisms*, Macmillan.

LIVINGSTON, W. K. (1948), 'The vicious circle in causalgia', *Ann. N.Y. Acad. Sci.*, vol. 50, p. 247.

LIVINGSTON, W. K. (1953), 'What is pain?', *Sci. Amer.*, vol. 196 (March), p. 59.

LORENTE DE NÓ, R. (1938), 'Analysis of the activity of the chains of internuncial neurons', *J. Neurophysiol.*, vol. 1, p. 207.

MacCarty, C. S., and Drake, R. L. (1956), 'Neurosurgical procedures for the control of pain', *Proc. Staff Meetings Mayo Clin.*, vol. 31, p. 208.

MacLean, P. (1958), 'Psychosomatics', *Handbook of Physiology*, vol. 3, p. 1723.

Mark, V. H., Ervin, F. R., and Yakovlev, P. I. (1963), 'Stereotactic thalamotomy', *Arch. Neurol.*, vol. 8, p. 528.

Marshall, H. R. (1894), *Pain, Pleasure, and Aesthetics*, Macmillan.

Mayer, D. J., Wolfle, T. L., Akil, H., Carder, B., and Liebeskind, J. C. (1971), 'Analgesia from electrical stimulation in the brainstem of the rat', *Science*, vol. 174, p. 1351.

McMurray, G. A. (1950), 'Experimental study of a case of insensitivity to pain', *Arch. Neurol. Psychiat.*, vol. 64, p. 650.

Melzack, R. (1961), 'The perception of pain', *Sci. Amer.*, vol. 204, (February), p. 41.

Melzack, R. (1965), 'Effects of early experience on behaviour: experimental and conceptual considerations', in P. H. Hoch and J. Zubin (eds.), *Psychopathology of Perception*, Grune & Stratton.

Melzack, R. (1969), 'The role of early experience in emotional arousal', *Ann. N.Y. Acad. Sci.*, vol. 159, p. 721.

Melzack, R. (1971), 'Phantom limb pain: implications for treatment of pathological pain', *Anaesthesiology*, vol. 35, p. 409.

Melzack, R. (1972), 'Mechanisms of pathological pain', in M. Critchley (ed.), *The Scientific Foundations of Neurology*, Heinemann.

Melzack, R. (1973), 'A questionnaire for the measurement of pain', manuscript in preparation.

Melzack, R., and Bridges, J. A. (1971), 'Dorsal column contributions to motor behaviour', *Exper. Neurol.*, vol. 33, p. 53.

Melzack, R., and Bromage, P. (1972), unpublished observations.

Melzack, R., and Casey, K. L. (1968), 'Sensory, motivational, and central control determinants of pain: a new conceptual model', in D. Kenshalo (ed.), *The Skin Senses*, C. C. Thomas.

Melzack, R., and Eisenberg, H. (1968), 'Skin sensory afterglows', *Science*, vol. 159, p. 445.

Melzack, R., Konrad, K., and Dubrovsky, B. (1968), 'Prolonged changes in visual system activity produced by somatic stimulation', *Exper. Neurol.*, vol. 20, p. 443.

Melzack, R., Konrad, K., and Dubrovsky, B. (1969), 'Prolonged changes in central nervous system activity produced by somatic and reticular stimulation', *Exper. Neurol.*, vol. 20, p. 416.

Melzack, R., Rose, G., and McGinty, D. (1962), 'Skin sensitivity to thermal stimuli', *Exper. Neurol.*, vol. 6, p. 300.

Melzack, R., and Schecter, B. (1965), 'Itch and vibration', *Science*, vol. 147, p. 1047.

216 References

MELZACK, R., and SCOTT, T. H. (1957), 'The effects of early experience on the response to pain', *J. comp. physiol. Psychol.*, vol. 50, p. 155.

MELZACK, R., STOTLER, W. A., and LIVINGSTON, W. K. (1958), 'Effects of discrete brainstem lesions in cats on perception of noxious stimulation', *J. Neurophysiol.*, vol. 21, p. 353.

MELZACK, R., and TORGERSON, W. S. (1971), 'On the language of pain', *Anesthesiology*, vol. 34, p. 50.

MELZACK, R., and WALL, P. D. (1962), 'On the nature of cutaneous sensory mechanisms', *Brain*, vol. 85, p. 331.

MELZACK, R., and WALL, P. D. (1965), 'Pain mechanisms: a new theory', *Science*, vol. 150, p. 971.

MELZACK, R., and WALL, P. D. (1970), 'Psychophysiology of pain', *Internat. Anesthesiol. Clinics*, vol. 8, p. 3.

MELZACK, R., WALL, P. D., and WEISZ, A. Z. (1963), 'Masking and metacontrast phenomena in the skin sensory system', *Exper. Neurol.*, vol. 8, p. 35.

MELZACK, R., WEISZ, A. Z., and SPRAGUE, L. T. (1963), 'Stratagems for controlling pain: contributions of auditory stimulation and suggestion', *Exper. Neurol.*, vol. 8, p. 239.

MENDELL, L. M., and WALL, P. D. (1965), 'Presynaptic hyperpolarization: a role for fine afferent fibres', *J. Physiol*, vol. 172, p. 274.

MERSKEY, H., and SPEAR, F. H. (1967), *Pain: Psychological and Psychiatric Aspects*, Baillière, Tindall & Cassell.

MEYER, G. A., and FIELDS, H. L. (1972), 'Causalgia treated by selective large fibre stimulation of peripheral nerve', *Brain*, vol. 95, p. 163.

MILNER, P. M. (1970), *Physiological Psychology*, Holt, Rinehart & Winston.

MITCHELL, S. W. (1872), *Injuries of Nerves and their Consequences*, Lippincott.

MOORE, B. (1946), 'Pain in an amputation stump associated with spinal anesthesia', *Med. J. Austral.*, vol. 2, p. 645.

MORGAN, C. L. (1961), *Introduction to Psychology*, McGraw-Hill.

MÜLLER, J. (1842), *Elements of Physiology*, Taylor.

NAFE, J. P. (1934), 'The pressure, pain and temperature senses', in C. A. Murchison (ed.), *Handbook of General Experimental Psychology*, Clark University Press.

NAKAHAMA, H., NISHIOKA, S., and OTSUKA, T. (1966), 'Excitation and inhibition in ventrobasal thalamic neurons before and after cutaneous input deprivation', *Progr. Brain Res.*, vol. 21, p. 180.

NASHOLD, B. S., and FRIEDMAN, H. (1972), 'Dorsal column stimulation for pain: a preliminary report on 30 patients', *J. Neurosurg.*, vol. 36, p. 590.

NATHAN, P. W. (1956), 'Reference of sensation at the spinal level', *J. Neurol. Neurosurg. Psychiat.*, vol. 19, p. 88.

NATHAN, P. W. (1962), 'Pain traces left in the central nervous system', in C. A. Keele and R. Smith (eds.), *The Assessment of Pain in Man and Animals*, Livingstone.

NATHAN, P. W. (1963), 'Results of antero-lateral cordotomy for pain in cancer', *J. Neurol. Neurosurg. Psychiat.*, vol. 26, p. 353.

NAUNYN, B. (1889), 'Über die Auslösung von Schmerzempfindung durch Summation sich zeitlich folgender sensibelen Erregungen', *Arch. exper. Pathol. Pharmakol.*, vol. 25, p. 272.

NAUTA, W. J. H. (1958), 'Hippocampal projections and related neural pathways to the midbrain in the cat', *Brain*, vol. 81, p. 319.

NEFF, W. D. (1961), 'Neural mechanisms of auditory discrimination', in W. A. Rosenblith (ed.), *Sensory Communication*, Wiley.

NOORDENBOS, W. (1959), *Pain*, Elsevier Press.

OLDS, M. E., and OLDS, J. (1962), 'Approach–escape interactions in the rat brain', *Amer. J. Physiol.*, vol. 203, p. 803.

OLDS, M. E., and OLDS, J. (1963), 'Approach–avoidance analysis of rat diencephalon', *J. comp. Neurol.*, vol. 120, p. 259.

PAPEZ, J. W., and STOTLER, W. A. (1940), 'Connections of the red nucleus', *Arch. Neurol. Psychiat.*, vol. 44, p. 776.

PARSONS, C. M., and GOETZL, F. R. (1945), 'Effect of induced pain on pain threshold', *Proc. Soc. Exper. Biol.*, vol. 60, p. 327.

PAVLOV, I. P. (1927), *Conditioned Reflexes*, Milford.

PAVLOV, I. P. (1928), *Lectures on Conditioned Reflexes*, International Publishers.

PEARSON, A. A. (1952), 'Role of gelatinous substance of spinal cord in conduction of pain', *Arch. Neurol. Psychiat.*, vol. 68, p. 515.

PERL, E. R. (1971), 'Is pain a specific sensation?', *J. psychiat. Res.*, vol. 8, p. 273.

POMERANZ, B., WALL, P. D., and WEBER, W. V. (1968), 'Cord cells responding to fine myelinated afferents from viscera, muscle and skin', *J. Physiol.*, vol. 199, p. 511.

REYNOLDS, D. V. (1969), 'Surgery in the rat during electrical analgesia induced by focal brain stimulation', *Science*, vol. 164, p. 444.

REYNOLDS, D. V. (1970), 'Reduced response to aversive stimuli during focal brain stimulation: electrical analgesia and electrical anesthesia', in D. V. Reynolds and A. E. Sjoberg (eds.), *Neuroelectric Research*, C. C. Thomas.

REYNOLDS, O. E., and HUTCHINS, H. C. (1948), 'Reduction of central hyper-irritability following block anesthesia of peripheral nerve', *Amer. J. Physiol.*, vol. 152, p. 658.

ROBERTS, W. W. (1962), 'Fear-like behaviour elicited from dorsomedial thalamus of cat', *J. comp. physiol. Psychol.*, vol. 55, p. 191.

Rose, J. E., and Mountcastle, V. B. (1959), 'Touch and kinesthesis', *Handbook of Physiology*, vol. 1, p. 387.

Rossi, G. F., and Zanchetti, A. (1957), 'The brainstem reticular formation', *Arch. italiennes de Biologie*, vol. 95, p. 199.

Rothman, S. (1943), 'The nature of itching', *Res. Publ. Assn nerv. ment. Dis.*, vol. 23, p. 110.

Rubins, J. L., and Friedman, E. D. (1948), 'Asymbolia for pain', *Arch. Neurol. Psychiat.*, vol. 60, p. 554.

Ruffini, A. (1905), 'Les dispositifs anatomiques de la sensibilité cutanée sur les expansions nerveuses de la peau chez l'homme et quelques autres mammifères', *Rev. gen. Histol.*, vol. 1, p. 421.

Russell, W. R., and Spalding, J. M. K. (1950), 'Treatment of painful amputation stumps', *Brit. med. J.*, vol. 2, p. 68.

Satoh, M., and Takagi, H. (1971), 'Enhancement by morphine of the central descending inhibitory influence on spinal sensory transmission', *Eur. J. Pharmacol.*, vol. 14, p. 60.

Schreiner, L., and Kling, A. (1953), 'Behavioural changes following rhinencephalic injury in cat', *J. Neurophysiol.*, vol. 16, p. 643.

Selzer, M., and Spencer, W. A. (1969), 'Convergence of visceral and cutaneous afferent pathways in the lumbar spinal cord', *Brain Res.*, vol. 14, p. 331.

Semmes, J., and Mishkin, M. (1965), 'Somatosensory loss in monkeys after ipsilateral cortical ablation', *J. Neurophysiol.*, vol. 28, p. 473.

Shealy, C. N., Mortimer, J. T., and Hagfors, N. R. (1970), 'Dorsal column electroanalgesia', *J. Neurosurg.*, vol. 32, p. 560.

Sherrington, C. S. (1900), 'Cutaneous sensations', in E. A. Schäfer (ed.), *Textbook of Physiology*, Pentland.

Sherrington, C. S. (1906), *Integrative Action of the Nervous System*, Scribner.

Shimazu, H., Yanagisawa, N., and Garoutte, B. (1965), 'Corticopyramidal influences on thalamic somatosensory transmission in the cat', *Jap. J. Physiol.*, vol. 15, p. 101.

Signer, E., and Galston, A. W. (1972), 'Education and science in China', *Science*, vol. 175, p. 15.

Simmel, M. L. (1956), 'On phantom limbs', *A.M.A. Arch. Neurol. Psychiat.*, vol. 75, p. 637.

Simmel, M. L. (1958), 'The conditions of occurrence of phantom limbs', *Proc. Amer. Phil. Soc.*, vol. 102, p. 492.

Sinclair, D. C. (1955), 'Cutaneous sensation and the doctrine of specific nerve energies', *Brain*, vol. 78, p. 584.

Sinclair, D. C. (1967), *Cutaneous Sensation*, Oxford University Press.

SPENCER, W. A., and APRIL, R. S. (1970), 'Plastic properties of monosynaptic pathways in mammals', in G. Horn and R. A. Hinde (eds.), *Short-term Changes in Neural Activity and Behaviour*, Cambridge University Press.

SPIEGEL, E. A., KLETZKIN, M., and SZEKELEY, E. G. (1954) 'Pain reactions upon stimulation of the tectum mesencephali', *J. Neuropath. exper. Neurol.*, vol. 13, p. 212.

SPIEGEL, E. A., and WYCIS, H. T. (1966), 'Present status of stereoencephalotomies for pain relief', *Confinia Neurologica*, vol. 27, p. 7.

STERNBACH, R. A. (1968), *Pain: A Psychophysiological Analysis*, Academic Press.

STERNBACH, R. A. (1970), 'Strategies and tactics in the treatment of patients with pain', in B. L. Crue (ed.), *Pain and Suffering: Selected Aspects*, C. C. Thomas.

STERNBACH, R. A., and TURSKY, B. (1964), 'On the psychophysical power function in electric shock', *Psychosom. Sci.,* vol. 1, p. 217.

STERNBACH, R. A., and TURSKY, B. (1965), 'Ethnic differences among housewives in psychophysical and skin potential responses to electric shock', *Psychophysiology*, vol. 1, p. 241.

STEVENS, S. S., CARTON, A. S., and SHICKMAN, G. M. (1958), 'A scale of apparent intensity of electric shock', *J. exper. Psychol.*, vol. 56, p. 328.

SUNDERLAND, S. (1968), *Nerves and Nerve Injuries*, E. and S. Livingstone.

SUNDERLAND, S., and KELLY, M. (1948), 'The painful sequelae of injuries to peripheral nerves', *Aust. N.Z. J. Surg.*, vol. 18, p. 75.

SWEET, W. H. (1959), 'Pain', *Handbook of Physiology*, vol. 1, p. 459.

SZASZ, T. S. (1968), 'The psychology of persistent pain: a portrait of l'homme douloureux', in A. Soulairac, J. Cahn and J. Charpentier (eds.), *Pain*, Academic Press.

SZENTAGOTHAI, J. (1964), 'Neuronal and synaptic arrangement in the substantia gelatinosa Rolandi', *J. comp. Neurol.*, vol. 122, p. 219.

TAUB, A. (1964), 'Local, segmental and supraspinal interaction with a dorsolateral spinal cutaneous afferent system', *Exper. Neurol.*, vol. 10, p. 357.

TITCHENER, E. B. (1909–10), *A Textbook of Psychology*, Macmillan.

TITCHENER, E. B. (1920), 'Notes from the psychological laboratory of Cornell University', *Amer. J. Psychol.*, vol. 31, p. 212.

TOWER, S. S. (1943), 'Pain: definition and properties of the unit for sensory reception', *Res. Publ. Assn nerv. ment. Dis.*, vol. 23, p. 16.

TRAVELL, J., and RINZLER, S. H. (1946), 'Relief of cardiac pain by local block of somatic trigger areas', *Proc. Soc. Exper. Biol. Med.*, vol. 63, p. 480.

TRAVELL, J., and RINZLER, S. H. (1952), 'The myofascial genesis of pain', *Postgrad. Med.*, vol. 11, p. 425.

VERHAART, W. J. C. (1949), 'The central tegmental tract', *J. comp. Neurol.*, vol. 90, p. 173.

WAGMAN, I. H., and PRICE, D. D. (1969), 'Responses of dorsal horn cells of *M. mulatta* to cutaneous and sural nerve A and C fibre stimuli', *J. Neurophysiol.*, vol. 32, p. 803.

WALL, P. D. (1960), 'Cord cells responding to touch, damage and temperature of the skin', *J. Neurophysiol.*, vol. 23, p. 197.

WALL, P. D. (1961), 'Two transmission systems for skin sensations', in W. A. Rosenblith (ed.), *Sensory Communication*, Wiley.

WALL, P. D. (1964), 'Presynaptic control of impulses at the first central synapse in the cutaneous pathway', *Progr. Brain Res.*, vol. 12, p. 92.

WALL, P. D. (1970), 'The sensory and motor role of impulses travelling in the dorsal columns towards cerebral cortex', *Brain*, vol. 93, p. 505.

WALL, P. D., and CRONLY-DILLON, J. R. (1960), 'Pain, itch and vibration', *Arch. Neurol.*, vol. 2, p. 365.

WALL, P. D., and SWEET, W. H. (1967), 'Temporary abolition of pain', *Science*, vol. 155, p. 108.

WALTERS, A. (1961), 'Psychogenic regional pain alias hysterical pain', *Brain*, vol. 84, p. 1.

WARD, A. A. (1969), 'The epileptic neuron: chronic foci in animals and man', in H. H. Jasper, A. A. Ward and A. Pope (eds.), *Basic Mechanisms of the Epilepsies*, Little, Brown.

WEDDELL, G. (1955), 'Somesthesis and the chemical senses', *Annu. Rev. Psychol.*, vol. 6, p. 119.

WEISKRANTZ, L. (1963), 'Contour discrimination in a young monkey with striate cortex ablation', *Neuropsychologia*, vol. 1, p. 145.

WHITE, J. C., and SWEET, W. H. (1969), *Pain and the Neurosurgeon*, C. C. Thomas.

WISSLER, C. (1921), 'The sun dance of the Blackfoot Indians', *Amer. Mus. Nat. Hist.*, *Anthropology Papers*, vol. 16, pp. 223–70.

YAHR, M. D., and PURPURA, D. P. (Eds.) (1967), *Neurophysiological Basis of Normal and Abnormal Motor Activities*, Raven Press.

ZBOROWSKI, M. (1952), 'Cultural components in responses to pain', *J. soc. Issues*, vol. 8, p. 16.

ZOTTERMAN, Y. (1959), 'Thermal sensations', *Handbook of Physiology*, vol. 1, p. 431.

Index

DATE DUE
